W9-BNX-027

SELF-PRODUCING SYSTEMS
Implications and Applications of Autopoiesis

Contemporary Systems Thinking

Series Editor: Robert L. Flood
University of Hull
Hull, United Kingdom

LIBERATING SYSTEMS THEORY
Robert L. Flood

OPERATIONAL RESEARCH AND SYSTEMS
The Systemic Nature of Operational Research
Paul Keys

SELF-PRODUCING SYSTEMS
Implications and Applications of Autopoiesis
John Mingers

SYSTEMS METHODOLOGY FOR THE MANAGEMENT SCIENCES
Michael C. Jackson

A Continuation Order Plan is available for this series. A continuation order will bring delivery of each new volume immediately upon publication. Volumes are billed only upon actual shipment. For further information please contact the publisher.

SELF-PRODUCING SYSTEMS
Implications and Applications of Autopoiesis

John Mingers

University of Warwick
Coventry, England

PLENUM PRESS • NEW YORK AND LONDON

Library of Congress Cataloging-in-Publication Data

On file

ISBN 0-306-44797-5

© 1995 Plenum Press, New York
A Division of Plenum Publishing Corporation
233 Spring Street, New York, N. Y. 10013

Printed in the United States of America

For my daughters Laura and Emma
who enabled me to observe
the development of a consensual domain
and for Julie who co-produced them.

Foreword

John Mingers' new volume, *Self-Producing Systems: Implications and Applications of Autopoiesis,* is a much-needed reference on autopoiesis, a subject penetrating many disciplines today. I can genuinely say that I enjoyed reading the book as it took me stage by stage through a clear and easy-to-grasp understanding of the concepts and ideas of autopoiesis and then, as the book's title suggests, on through their applications. I found the summary in Chapter 12 particularly useful, helping to crystalize the main points of each chapter. The book conveyed enthusiasm for the subject and stimulated my interest in it. At times the book is demanding, but only because of the breadth of the subject matter, the terms and concepts associated with its parts, and the challenge of keeping hold of all this in the mind at once. This is an exceptional text.

ROBERT L. FLOOD

Hull, UK

Preface

In recent years Maturana's and Varela's concept of autopoiesis, originally a biological concept, has made a remarkable impact not just on a single area, but across widely differing disciplines such as sociology, policy science, psychotherapy, cognitive science, and law. Put very briefly, the term *autopoiesis* connotes the idea that certain types of systems exist in a particular manner—they are *self-producing* systems. In their operations they continuously produce their own constituents, their own components, which then participate in these same production processes. Such an autopoietic system has a circular organization, which closes in on itself, its outputs becoming its own inputs. This gives it an important degree of independence or autonomy from its environment since its own operations ensure, within limits, its future continuation. Maturana and Varela contend that all living systems are autopoietic and that autopoiesis explains their particular characteristics.

It is interesting, however, that autopoiesis has had more impact elsewhere than in its own original domains—biology and neurophysiology—where interest is only just beginning. It is fascinating that a single concept can be so stimulating in such diverse fields. However, this also poses a problem: disciplines tend to be self-referring and insulated from one another, so the appropriation of autopoiesis in different domains has often been strikingly different. It has also sometimes been based on very partial, if not unsound, readings of the original ideas. So far, there has been no single work that presents both the ideas in themselves and their applications across the spectrum of subjects. That is the intention of this book.

The book should be seen not as a summation and evaluation of autopoiesis—it is much too early for that. Rather, it is intended as an opening-up of autopoiesis, as a facilitation of even more productive and well-founded work. Autopoiesis requires opening-up in a number of ways:

- The original language of autopoiesis is opaque and convoluted and in a sense closed. It is hard to penetrate without much effort.

I therefore try to re-export the ideas in more transparent language.

- The theoretical work is also closed in that it makes almost no contact or reference with other bodies of ideas. Maturana, especially, may claim that it is radically new, but on examination it is considering some of the age-old questions. I therefore try whenever possible to point out possible connections and resonances with other areas.

- The work has been taken up in many different disciplines, but each has its own interpretations, its own concerns, and its own applications. I think it is useful to bring each of these to the attention of those working in the others.

The book is, therefore, largely expository rather than evaluative or critical. However, I have adopted a more critical stance in certain sections, such as those on philosophy and social theory, and I have indicated areas of debate in the final chapter.

The book is intended for a transdisciplinary audience, an audience either of people who are interested in autopoiesis but have found it unrewardingly difficult to get into or of those who may consider themselves very knowledgeable about autopoiesis in their own disciplines but are interested in finding out about other areas of application. My primary intention has always been to be as clear as possible about the underlying ideas and to explicate the rather bare, neologistic language. I have, however, used the original terminology because it is, once understood, very precise, and to provide a gateway into the original texts, which I would strongly recommend. I have tried to keep the chapters on particular disciplines fairly self-contained, but, clearly, I have not repeated the background ideas each time, and an initial study of the early chapters is important. I have also tried to be as thorough as I can in detailing the wide range of references to autopoiesis, regarding this as a resource for the reader. This does mean that the most recent works, which appeared during the writing of the book, have been mentioned but not fully assimilated. There is a brief guide to the primary literature in the bibliography.

I first came across autopoiesis in 1976 and was both fascinated and frustrated by the early papers. It has taken many years for me to gain whatever clarity I have, and yet I still find many puzzling features, so I feel that it is important to keep working at it and not expect instant enlightenment. It is worth it. Some of the groundwork for this book has been covered in three papers of mine (Mingers, 1989a, 1990, 1991) published by Plenum Press in the journal *Systems Practice* and one (Mingers, 1992b) published in the *International Journal of General Systems*.

Many people have helped me by commenting on various drafts. I am particularly grateful to Roger Harnden, who read it all and made valuable comments extremely speedily, and Gail Fleischaker, who was exceptionally thorough and helpful. Other readers were David Johnson, Francisco Varela, Bebe Speed, Mike Power, Bob Cooper, Peter Binns, Andy Clark, and John Pickering. They will inevitably not agree with my response to their comments, but it is my book, not theirs, and I must take full responsibility for its flaws.

I am also indebted to Zoe Grimsdale for the diagrams.

Contents

The Development of Autopoiesis

Physis also, the arising of something from out of itself, is a bringing-forth, <u>poiesis</u>.
Physis is indeed poiesis in the highest sense. For what presences by means of physis
has the bursting open belonging to bringing-forth, e.g., the bursting open of a
blossom into bloom, in itself (<u>en heautoi</u>). In contrast, what is brought forth by the
artisan or the artist, e.g., <u>the silver chalice</u>, has the bursting open belonging to
bringing-forth, not in itself, but in another (<u>en alloi</u>), in the craftsman or artist.
Heidegger (1977, p. 293, my emphases). Published originally in German
in 1954.

1.1. Introduction

What is the nature of life? What distinguishes living systems from those
that appear equally complex but we do not call living? What is the basic
unit of biology—is it the species, or the gene, or the individual? What is
the nature of cognition? Is it pure abstract thought, or is it intimately
connected to our bodily structures? Are our cognitions, our descriptions
of the world, reflections of an independent reality or constructions of
ourselves, the observer? Is there an independent reality at all, and if
there is can we interact with it? What is the nature of social reality? Are
we unwitting participants in supraindividual systems that are autonomous and beyond our control? How can we deal with self-reference and
the contradictions it appears to create?

These are some of the many fundamental questions addressed by
autopoiesis, the creation of Humberto Maturana and Francisco Varela.
Formulated in the early 1970s as an explanation for the nature of living
systems and presented in a highly abstract form, their work has, slowly
at first, generated an enormous amount of interest in a diverse set of
disciplines. It has triggered debates about the nature of family reality (in
psychotherapy), the ontology of law, the self-constitution of social systems, and the grounding of cognitive science and artificial intelligence.
This is on top of debates within its own domain—biology—about the
origin of life and artificial living systems. Some idea of this proliferation
will be given below.

Clearly, autopoiesis has been extremely productive—why is this? First, I think, because autopoiesis addresses major themes and does so just at a time when they have become the preoccupation of many disciplines. The loss of the certainties of modernism, initially from within modernism itself—in philosophy and sociology—and, latterly, from postmodernism, have brought basic questions of epistemology, ontology, and language to the fore. Second, because autopoiesis addresses these themes in an original and exciting way, turning traditional philosophical problems such as autonomy, self-reference, and the nature of mind on their heads. Third, because the work as a whole has a consistency and coherence across a wide range of domains that is rare indeed.

This is not to prejudge the final significance of autopoiesis. Although I feel confident about the basic biological formulation, application to other domains has been characterized by tremendous controversy and has generally led to serious questioning of the discipline's philosophy. This itself may, in the end, prove the greatest contribution of autopoiesis to other disciplines, but such judgments remain for the future. In the present, autopoiesis is often seen as an esoteric and rather mysterious subject, open to only a few initiates. At the same time, a significant number of people using autopoiesis do so with a hazy, if not simply incorrect, understanding. It is the primary purpose of this book to try to alleviate both difficulties by explaining, as simply as possible, the underlying ideas and language of the field and by exploring the various applications and implications of autopoiesis.

That said, it is still important to study the original papers of Maturana and Varela to appreciate the originality and beauty of their ideas, but initially these make very difficult reading. Their style makes little concession to the reader. It is dense with ideas expressed with almost mathematical sparseness and uses many common words—e.g., *unity, structure, organization*—in very precise but uncommon ways. Some guidance in using the source literature is given at the start of the bibliography.

1.2. *Analysis of Citations*

An analysis of the number of papers and books citing Maturana's and Varela's work (excluding self-citation) is shown in Fig. 1.1. Each work is counted only once even though it may cite a number of papers, so this is not a record of all citations. This analysis has been produced primarily from the works referred to in this book, and as such it must be seen simply as a sample. Still, it must represent a high proportion of all works citing Maturana and Varela, and I am not aware of any particular

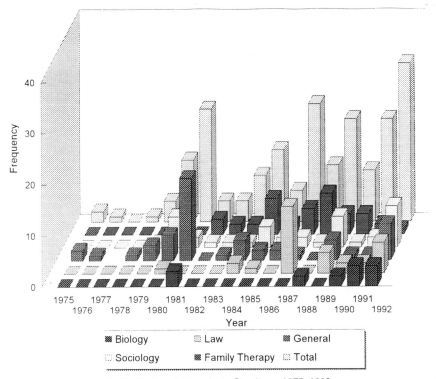

FIGURE 1.1. Autopoiesis Citations, 1975–1992.

biases apart from the exclusion of non-English papers. These are likely to be largely in German and in sociology and law, following Luhmann's work.

The chart well documents the rise in interest in autopoiesis, the total rising from four or five citations per year in the late seventies to 20 or 30 per year in the early nineties. Particular peaks are generally attributable to whole books of papers or special issues of journals in certain years. Within the various disciplines, family therapy has the highest overall number, with interest most intense during the late eighties. Most other subjects show a general rise recently, with 1992 being a particularly strong year.

1.3. Overview of This Book

This book is divided into three main sections: autopoiesis in the physical domain, theories of cognition, and applications within particular disciplines. In Part I, Chapters 2 and 3 develop the essential idea of auto-poiesis as an explanation of living systems and the implications of this idea in general and for biology in particular, including a debate about whether there could be nonmolecular living systems. Chapter 4 illustrates various formalisms relevant to autopoiesis, in particular Varela's extension of Spencer Brown's "Laws of Form," and a computer model of autopoiesis.

In Part II, Chapters 5 and 6 develop the cognitive theories of the early papers, showing how the evolution of a nervous system generates new domains of interaction culminating in language, description, and the observer. Chapter 7 discusses the philosophical implications of the original work and Maturana's more recent development of radical constructivism. It includes a comparison of Maturana with phenomenologists such as Husserl and Heidegger, and concludes with a critical interpretation of his position through Bhaskar's critical realism.

Part III consists of chapters about the application of autopoiesis to various disciplines. Chapter 8 covers sociology and organization theory, including discussion of Maturana's own social theory, a comparison with Giddens' structuration theory, and a detailed analysis of Luhmann's theory of society as autopoietic communication. Chapter 9 looks at the development of Luhmann's interpretation of law and the debates this has engendered. Chapter 10 considers the impact of Maturana's later theories on family therapy, where constructivism has again led to serious debate. Chapter 11 focuses on Varela's recent development of the cognitive theories into a view of cognition as *embodied* and *enactive* and the relationship of this view to cognitive science, artificial intelligence, and computing.

The book concludes (Chapter 12) with an appreciation of the importance of autopoiesis and a summary of the main points of debate.

1.4.   Conclusion

Whatever the final judgment on autopoiesis, I believe that it will stand as *an* example, if not *the* example, of outstanding work in the field of systems *per se*. It is founded on genuine knowledge of the relevant domains—biology and neurophysiology—but this is molded within a strong systems perspective, which supplies genuinely new insights. There is little new empirical work, rather a reconceptualization of what already exists together with a willingness to pursue implications rigorously, even though they may lead to unconventional conclusions. Autopoiesis transcends a common systems distinction between hard and soft, beginning, as it does, with natural science, yet generating an explanation for the interpretive and hermeneutic characteristics of human beings and their "languaging."

I

Autopoiesis In The Physical Domain

The Organization of Living Systems

It became clear that one, perhaps the, major function of the living cell was the constant re-creation of itself from within.
Rose (1970, p. 78)

2.1. The Essential Idea of Autopoiesis

The fundamental question Maturana and Varela set out to answer is: what distinguishes entities or systems that we would call living from other systems, apparently equally complex, which we would not? How, for example, should a Martian distinguish between a horse and a car? This is an example that Monod (1974, p. 19) uses in addressing the similar but not identical question of distinguishing between natural and artificial systems.

This has always been a problem for biologists, who have developed a variety of answers. First came vitalism (Bergson, 1911; Driesch, 1908), which held that there is some substance or force or principle, as yet unobserved, which must account for the peculiar characteristics of life. Then systems theory, with the development of concepts such as feedback, homeostasis, and open systems, paved the way for explanations of the complex, goal-seeking behavior of organisms in purely mechanistic terms (for example, Cannon, 1939; Priban, 1968). While this was a significant advance, such mechanisms could equally well be built into simple machines that would never qualify as living organisms.

A third approach, the most common recently, is to specify a list of necessary characteristics that any living organism must have—such as reproductive ability, information-processing capabilities, carbon-based chemistry, and nucleic acids (see, for example, Miller, 1978; Bunge, 1979). The first difficulty with this approach is that it is entirely descriptive and not in any real sense explanatory. It works by observing systems that are accepted as living and noting some of their common characteristics. However, this tactic assumes precisely that which is in need of explanation—the distinction between the living and the nonliving. The approach fails to define the characteristics particular to living systems alone or to give any explanation as to how such characteristics might generate the observed phenomena. Second, there is, inevitably, always a lack of agreement about the contents of such lists. Any two lists will contain different characteristics, and it is difficult to prove that every feature in a list is really necessary or that the list is actually complete.

Maturana's and Varela's work is based on a number of fundamental observations about the nature of living systems. They will be introduced briefly here but discussed in more detail in later chapters.

1. Somewhat in opposition to current trends that focus on the species or the genes (Dawkins, 1978), Maturana and Varela pick out the single, biological individual (for instance, a single celled creature such as an amoeba) as the central example of a living system. One essential feature of such living entities is their individual autonomy. Although they are part of organisms, populations, and species and are affected by their environment, individuals are bounded, self-defined entities.

2. Living systems operate in an essentially mechanistic way. They consist of particular components that have various properties and interactions. The overall behavior of the whole is generated purely by these components and their properties through the interactions of neighboring elements. Thus any explanation of living systems must be a purely mechanistic one.

3. All explanations or descriptions are made by observers (i.e., people) who are external to the system. One must not confuse that which pertains to the observer with that which pertains to the observed. Observers can perceive both an entity and its environment and see how the two relate to each other. Components within an entity, however, cannot do this, but act purely in response to other components.

4. The last two lead to the idea that any explanation of living systems should be nonteleological, i.e., it should not have recourse to ideas of function and purpose. The observable phenomena of living systems result purely from the interactions of neighboring internal components. The observation that certain parts appear to have a function with regard to the whole can be made only by

an observer who can interact with both the component and with the whole and describe the relation of the two.

To explain the nature of living systems, Maturana and Varela focus on a single, basic example—the individual, living cell. Briefly, a cell consists of a cell membrane or boundary enclosing various structures such as the nucleus, mitochondria, and lysosomes, as well as many (and often complex) molecules produced from within. These structures are in constant chemical interplay both with each other and, in the case of the membrane, with their external medium. It is a dynamic, integrated chemical network of incredible sophistication (see for example Alberts *et al.*, 1989; Freifelder, 1983; Raven and Johnson, 1991).

What is it that characterizes this as an autonomous, dynamic, living whole? What distinguishes it from a machine such as a chemical factory which also consists of complex components and interacting processes of production forming an organized whole? It cannot be to do with any functions or purposes that any single cell might fulfill in a larger multicellular organism since there are single-celled organisms that survive by themselves. Nor can it be explained in a reductionist way through particular structures or components of the cell, such as the nucleus or DNA/RNA. The difference must stem from the way the parts are organized as a whole. To understand Maturana and Varela's answer, we need to look at two related questions—what is it that the cell does, that is, what is it that the *cell produces*? and what is it that *produces the cell*? By this I mean the cell itself rather than the results of its reproduction.

What does a cell do? This will be looked at in detail in Section 2.3 but, in essence, it produces many complex and simple substances which remain in the cell (because of the cell membrane) and participate in those very same production processes. Some molecules are excreted from the cell, through the membrane, as waste. What is it that produces the components of the cell? With the help of some basic chemicals imported from its medium, the cell produces its own constituents. So a cell produces its own components, which are therefore what produces it in a circular, ongoing process (Fig. 2.1).

It produces, and is produced by, nothing other than itself. This simple idea is all that is meant by *autopoiesis*. The word means "self-producing," and that is what the cell does: it continually produces itself. Living systems are autopoietic—they are organized in such a way that their processes produce the very components necessary for the continuance of these processes. Systems which do not produce themselves are called *allopoietic*, meaning "other-producing"—for example, a river or a crystal. Maturana and Varela also refer to human-created systems as *het-*

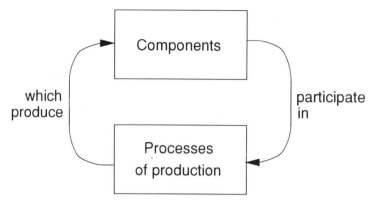

FIGURE 2.1. Circular Processes of Production.

eropoietic. An example is a chemical factory. Superficially, this is similar to a cell, but it produces chemicals that are used elsewhere, and is itself produced or maintained by other systems. It is not *self-producing.*

At first sight this may seem an almost trivial idea, yet further contemplation reveals how significant it is. For example:

1. Imagine trying to build an autopoietic machine. Save for energy and some basic chemicals, everything within it would itself have to be produced by the machine itself. So there would have to be machines to produce the various components. Of course, these machines themselves would have to be produced, maintained, and repaired by yet more machines, and so on, all within the same single entity. The machine would soon encompass the whole economy.

2. Suppose that you succeed. Then surely what you have created would be autonomous and independent. It would have the ability to construct and reconstruct itself, and would, in a very real sense, be no longer controlled by us, its creators. Would it not seem appropriate to call it living?

3. As life on earth originated from a sea of chemicals, a cell in which a set of chemicals interacted such that the cell created and re-created its own constituents would generate a stable, self-defined entity with a vastly enhanced chance of further development. This indeed is the basis for current research, to be described in Section 2.4.1.

4. What of death? If, for some reason, either internal or external, any part of the self-production process breaks down, then there is nothing else to produce the necessary components and the

whole process falls apart. Autopoiesis is all or nothing—all the processes must be working, or the system disintegrates.

This, then, is the central idea of autopoiesis: a living system is one organized in such a way that all its components and processes jointly produce those self-same components and processes, thus establishing an autonomous, self-producing entity. This concept has nearly been grasped by other biologists, as the quotation from Rose at the start of this chapter shows. But Maturana and Varela were the first to coin a word for this life-generating mechanism, to set out criteria for it (Varela *et al.*, 1974), and to explore its consequences in a rigorous way.

Considering the derivation of the word itself, Maturana explains (Maturana and Varela, 1980, p. xvii) that he had the main idea of a circular, self-referring organization without the term *autopoiesis*. In fact, *Biology of Cognition* (Maturana, 1970b), the first major exposition of the idea, does not use it. Maturana coined the term in relation to the distinction between *praxis* (the path of arms, or action) and *poiesis* (the path of letters, or creation). However, it is interesting to see how closely Maturana's usage of *auto-* and *allopoiesis* is actually foreshadowed by the German phenomenological philosopher Martin Heidegger. In the quotation at the start of Chapter 1, Heidegger uses the term *poiesis* as a bringing-forth and draws the contrast between the self-production (*heautoi*) of nature and the other-production (*alloi*) that humans do. Heidegger's relevance to Maturana's work will be considered further in Section 7.5.2.

2.2.   Formal Specification of Autopoiesis

Now that I have sketched the idea in general terms, this section will describe in more detail Maturana's and Varela's specification and vocabulary. [Maturana (1980a) is a reasonable introduction and includes a good glossary.]

We begin from the observation that all descriptions and explanations are made by observers who distinguish an entity or phenomenon from the general background. Such descriptions always depend in part on the choices and purposes of the observer and may or may not correspond to the actual domain of the observed entity. That which is distinguished by an observer, Maturana calls a *unity*, that is, a whole distinguished from a background. In making the distinction, the properties which specify the unity as a whole are established *by the observer*. For example, in calling something "a car," certain basic attributes or defining features (it is mobile, carries people, is steerable) are specified. An observer may go further and *analyze* a unity into components and their relations. There are different, equally valid, ways in which this can be done. The result will be a description of a composite unity of components and the organization which combines its components together into a whole.

Maturana and Varela draw an important distinction between the *organization* of a unity and its *structure:*

> [Organization] refers to the relations between components that define and specify a system as a composite unity of a particular class, and determine its properties as such a unity . . . by specifying a domain in which it can interact as an unanalyzable whole endowed with constitutive properties.
>
> [Structure] refers to the actual components and the actual relations that these must satisfy in their participation in the constitution of a given composite unity [and] determines the space in which it exists as a composite unity that can be perturbed through the interactions of its components, but the structure does not determine its properties as a unity.
> Maturana (1978a, p. 32)

The organization consists of the relations among components and the necessary properties of the components that characterize or define the unity in general as belonging to a particular type or class. This determines its properties *as a whole*. At its most simple, we can illustrate this distinction with the concept of a square. A square is defined in terms of the (spatial) relations between components—a figure with four equal sides, connected together at right angles. This is its organization. Any particular physically existing square is a particular structure that embodies these relations. Another example is a an airplane, which may be defined by describing necessary components such as wings, engines, controls, brakes, seating, and the relations between them allowing it to fly. If a unity has such an organization, then it may be identified as a plane since this particular organization would produce the properties we expect in a plane as a whole. Structure, on the other hand, describes the actual components and actual relations of a particular real example of any such entity, such as the Boeing 757 I board at the airport.

This is a rather unusual use of the term *structure* (Andrew, 1979). Generally, in the description of a system, structure is contrasted with process to refer to those parts of the system which change only slowly; structure and organization would be almost interchangeable. Here, however, structure refers to both the static and dynamic elements. The distinction between structure and organization is between the reality of an actual example and the abstract generality lying behind all such examples. This is strongly reminiscent of the philosophy of classic structuralism in which an empirical surface "structure" of events is related to an unobservable deep structure ("organization") of basic relationships which generate the surface.

An existing, composite unity, therefore, has both a structure and an organization. There are many different structures that can realize the same organization, and the structure will have many properties and relations not specified by the organization and essentially irrelevant to it—for example, the shape, color, size, and material of a particular airplane. Moreover, the structure can change or be changed without necessarily altering the organization. For example, as the plane ages, has new parts installed, and gets repainted it still maintains its identity as a plane because its underlying organization has not changed. Some changes, however, will not be compatible with the maintenance of the organization—for example, a crash which converts the plane into a wreck.

The essential distinction between organization and structure is between a whole and its parts. Only the plane as a whole can fly—this is its constitutive property as a unity, its organization. Its parts, however, can interact in their own domains depending on all their properties, but they do so only as individual components. Sucking in a bird can stop an engine; a short circuit can damage the controls. These are perturbations of the structure, which may affect the whole and lead to a loss of organization or which may be compensable, in which case the plane is still able to fly.

With this background, we can consider Maturana's and Varela's definition of autopoiesis. A unity is characterized by describing the organization that defines the unity as a member of a particular class, that is, which can be seen to generate the observed behavior of unities of that type. Maturana and Varela see living systems as being essentially characterized as dynamic and autonomous and hold that it is their self-production which leads to these qualities. Thus the organization of living systems is one of self-production—autopoiesis. Such an organization can, of course, be realized in infinitely many structures.

A more explicit definition of an autopoietic system is

> A dynamic system that is defined as a composite unity as a network of productions of components that,
> a) through their interactions recursively regenerate the network of productions that produced them, and
> b) realize this network as a unity in the space in which they exist by constituting and specifying its boundaries as surfaces of cleavage from the background through their preferential interactions within the network, is an autopoietic system.
> Maturana (1980b, p. 29)

The first part of this quotation details the general idea of a system of self-production, while the second specifies that the system must be actually realized in an entity that produces its own boundaries. This latter point, about producing boundaries, is particularly important when one attempts to apply autopoiesis to other domains, such as the social world, and is a recurring point of debate. Notice also that the definition does not specify that the realization must be a physical one, although in the case of a cell it clearly is. This leaves open the idea of some abstract autopoietic systems such as a set of concepts, a cellular automaton, or a process of communication. What might the boundaries of such a system be? And would we really want to call such a system "living"? Again, this is the subject of much debate—see Section 3.3.2.

This somewhat bare concept is further developed by considering the nature of such an organization. In particular, as an organization it will involve particular relations among components. These relations, in the case of a physical system, must be of three types according to Maturana and Varela (1973): *constitution, specification,* and *order*. Relations of constitution concern the physical topology of the system (say, a cell)—its three-dimensional geometry. For example, that it has a cell membrane, that components are particular distances from each other, that they are the required sizes and shapes. Relations of specification determine that the components produced by the various production processes are in fact the specific ones necessary for the continuation of autopoiesis. Finally, relations of order concern the dynamics of the processes—for example, that the appropriate amounts of various molecules are produced at the correct rate and at the correct time. Specific examples of these relations will be given later, but it can be seen that these correspond roughly to specifying the "where," "what," and "when" of the complex production processes occurring in the cell.

It might appear that this description of relations "necessary" for autopoiesis has a functionalist, teleological tone. This is not really the case, as Maturana and Varela strongly object to such explanations. It is simply that, if such components and relationships do occur, they give rise to electrochemical processes that themselves produce further components and processes of the right types and at the right rates to generate an autopoietic system. But there is no necessity to this; it is simply a combination that does, or does not, occur, just as a plant may, or may not, grow depending on the combination of water, light, and nutrients.

In an early attempt to make this abstract characterization more operational, a computer model of an autopoietic cellular automaton was developed together with a six-point key for identifying an autopoietic system (Varela *et al.*, 1974). The key is specified as follows (pp. 192–193):

i) Determine, through interactions, if the unity has identifiable boundaries. If the boundaries can be determined, proceed to 2. If not, the
entity is indescribable and we can say nothing.

ii) Determine if there are constitutive elements of the unity, that is, components of the unity. If these components can be described, proceed
to 3. If not, the unity is an unanalyzable whole and therefore not an
autopoietic system.

iii) Determine if the unity is a mechanistic system, that is, the component
properties are capable of satisfying certain relations that determine in
the unity the interactions and transformations of these components. If
this is the case, proceed to 4. If not, the unity is not an autopoietic
system.

iv) Determine if the components that constitute the boundaries of the
unity constitute these boundaries through preferential neighborhood
interactions and relations between themselves, as determined by their
properties in the space of their interactions. If this is not the case, you
do not have an autopoietic unity because you are determining its
boundaries, not the unity itself. If 4 is the case, however, proceed to
5.

v) Determine if the components of the boundaries of the unity are produced by the interactions of the components of the unity, either by
transformation of previously produced components, or by transformations and/or coupling of non-component elements that enter the
unity through its boundaries. If not, you do not have an autopoietic
unity; if yes proceed to 6.

vi) If all the other components of the unity are also produced by the interactions of its components as in 5, and if those which are not produced by the interactions of other components participate as necessary
permanent constitutive components in the production of other components, you have an autopoietic unity in the space in which its components exist. If this is not the case, and there are components in the
unity not produced by components of the unity as in 5, or if there are
components of the unity which do not participate in the production of
other components, you do not have an autopoietic unity.

The first three criteria are general, specifying that there is an identifiable entity with a clear boundary, that it can be analyzed into components, and that it operates mechanistically, i.e., its operation is
determined by the properties and relations of its components. The core
autopoietic ideas are specified in the last three points. These describe a
dynamic network of interacting processes of production (vi), contained
within and producing a boundary (v) that is maintained by the preferential interactions of its components (iv). The key notions, especially
when considering the extension of autopoiesis to nonphysical systems,
are the idea of production of components, and the necessity for a boundary constituted by produced components.

These key criteria will be applied to the cell in the next section.

2.3.   An Illustration of Autopoiesis in the Cell

This section will describe briefly embodiments of the autopoietic relations outlined above in the chemistry of the cell. Alberts *et al.* (1989) or
Freifelder (1983) are good introductions to molecular biology, as is Raven
and Johnson (1991) to the cell.

2.3.1.   Applying the Six Criteria

Zeleny and Hufford (1992a) analyze a typical cell with the six key points.
A schematic of two typical cells is shown in Fig. 2.2. One is a eukaryotic
cell, i.e., one that has a nucleus, and the other is a prokaryotic cell,
which does not.

1. The cell has an identifiable boundary formed by the plasma
    membrane. Thus, the cell is identifiable.
2. The cell has identifiable components such as the mitochondria,
    the nucleus, and the membranous network known as the endo-
    plasmic reticulum. Thus, the cell is analyzable.
3. The components have electrochemical properties that follow
    general physical laws determining the transformations and in-

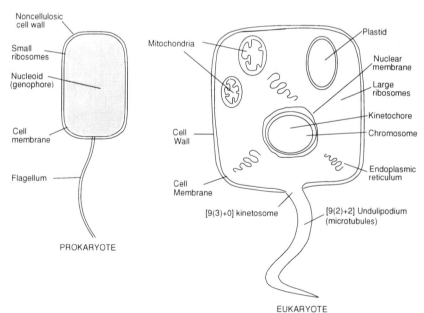

FIGURE 2.2. Prokaryote and Eukaryote Cells (after Margulis, 1993, p. 3).

teractions that occur within the cell. Thus, the cell is a mechanistic system.

4. The boundary of the cell is formed by a plasma membrane consisting of phospholipid molecules and certain proteins (Fig. 2.3). The lipid molecules are aligned in a double layer, forming a selectively permeable barrier; the proteins are wedged in this bilayer, mediating many of the membrane functions. A lipid molecule consists of two parts—a polar head, which is attracted to water, and a hydrocarbon (fatty) tail, which is repelled. In solution, the tails join together to form the two layers with the heads outside. The integral proteins also have areas that seek or avoid water. The boundary is therefore self-maintained through preferential neighborhood relations.

5. The lipid and protein components of the boundary are themselves produced by the cell. For example, most of the lipid molecules required for new membrane formation are produced by the endoplasmic reticulum, which is itself a complex, membranous component of the cell. The boundary components are thus self-produced.

6. All of the other components of the cell (e.g., the mitochondria, the nucleus, the ribosomes, the endoplasmic reticulum) are also produced by and within the cell. Certain chemicals (such as metal ions) not produced by the cell are imported through the

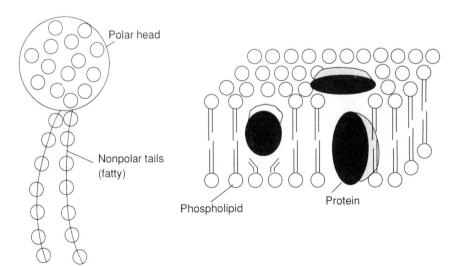

FIGURE 2.3. Phospholipid Membrane.

membrane and then become part of the operations of the cell. Cell components are thus self-produced.

Overall, therefore, a cell exhibits the autopoietic organization in the physical space of its molecular components.

2.3.2. *Autopoietic Relations of Constitution, Specification, and Order*

Apart from the six-point key, autopoiesis was also defined by three necessary types of relations. These can be illustrated as follows for a typical cell.

2.3.2.1. Relations of Constitution

Relations of constitution determine the three-dimensional shape and structure of the cell so as to enable the other relations of production to be maintained. This occurs through the production of molecules which, through their particular stereochemical properties, enable other processes to continue.

An obvious example is the construction of membranes or cell boundaries. In animal cells, the membrane surrounding the mitochondria, like that around the cell itself, serves to harbor cell contents and control the rate of reaction through diffusion. Various reactive molecules are distributed along the inner membrane in an appropriate order to allow energy-producing sequences to proceed efficiently. In plant cells, in addition to the plasma membrane, there is a cell wall, which consists of cellulose, a material made up of long, straight chains of glucose units packed together to form strong rigid threads. These give plants their rigidity.

A second example is the active sites on enzymatic proteins. These act as catalysts for most reactions, changing a particular substrate in an appropriate way to allow it to react more easily. Generally, the active site is found in certain specific parts of the enzyme molecule where the configuration of amino acids is structured to fit the particular substrate, sometimes with the help of "activators" or co-enzymes. The substrate molecule interlocks with the active site and in so doing changes appropriately so that it no longer fits, and thus frees itself.

2.3.2.2. Relations of Specification

These determine the identity, in chemical properties, of the components of the cell in such a way that through their interactions they participate in the production of the cell. There are two main types of structural correspondence, that among DNA, RNA, and the proteins they produce and that between enzymes and the substrates they catalyze.

Protein synthesis is particularly complex because each protein is formed by linking up to twenty different amino acids in a specific combination, often containing 300 or more units in all. This requires an RNA template molecule, tailor-made for each protein, containing specific spaces for each of the amino acids in order, together with an enzyme and *t*-RNA for each acid.

As already mentioned, enzymes are necessary to help most of the reactions in the cell, and again, each specific reaction requires an enzyme specific to the reaction and to the substrate involved. Hundreds of such enzymes are needed, and all must be produced by the cell.

2.3.2.3. Relations of Order

Relations of order concern the dynamics of the cell's production processes. Various chemicals and complex feedback loops ensure that both the rate and the sequence of the various production processes continue autopoiesis. For instance, the production of energy through oxidation is controlled by the amount of phosphate and ADP (adenosine diphosphate) in the mitochondria. At the same time, reactions that use energy actually produce ADP and phosphate so that, automatically, a high usage of energy leads to a high production rate of these necessary substances.

2.3.3.   Other Possible Autopoietic Systems

An interesting question leading from the idea of the cell as an autopoietic system is whether or not there are other instances of autopoietic systems. Are multicellular organisms also autopoietic systems? Maturana (1980a, p. 53) is equivocal, suggesting that organisms such as animals and plants may be second-order autopoietic systems, with the components being not the cells themselves but various molecules produced by the cells. On the other hand, he suggests that some cellular systems may not actually constitute autopoietic systems, but may be merely colonies. What about a system that appears to have a closed and circular organization but is not generally classified as living, such as the pilot light of a gas boiler? Finally, what about nonphysical systems such as the autopoietic automata mentioned in section 2.2.1 and described more fully in section 4.4, or systems such as a set of ideas or a society? These possibilities will be discussed in more detail in Section 3.3.

2.4. Applications of Autopoiesis in Biology and Chemistry

One would have expected that, given the importance and nature of its claims, autopoiesis would have had a major impact on the field of biology. In fact, for many years there was a noticeable reluctance to take the ideas seriously at all. In 1979, I wrote to an eminent British biologist—Professor Steven Rose at the Open University—querying the status of autopoiesis. He replied to the effect that he did not wish to comment on autopoiesis but that Maturana was a reputable biologist. One notable exception is Lynn Margulis (1993), whose own theory, that eukaryotic cells evolved through the symbiosis of simpler units, is itself quite controversial.

However, recently interest has been growing in two areas: research into the origins of life and the creation of chemical systems that, although not living, display some of the characteristics of autopoietic self-production. Autopoiesis has also been compared with Prigogine's (1980) dissipative structures (Briggs and Peat, 1985). Varela has also pursued work on the nature of the immune system, viewing it as organizationally closed but not autopoietic (Vaz and Varela, 1978; Varela et al., 1988; Varela and Coutinho, 1989, 1991; Varela and Anspach, 1991). However, as this topic is very technical and not of primary relevance, it cannot be pursued here.

2.4.1. Minimal Cells and the Origin of Life

There are two main lines of approach to theories concerning the origin of life on Earth (Fleischaker, 1990). In the first approach, based on study of the enzymes and genes, life is characterized as being molecular and a defining feature is the structure and function of the genes. In the second approach, life is characterized as cellular, and its defining feature is metabolic functioning within the cell. However, neither approach can really specify a standard or model for life against which important questions may be answered. In particular, at what point did prebiotic chemical systems become biotic living systems? And how could we recognize nonterrestrial living systems, which might be radically different in structure from our own (Dupuy and Varela, 1992)?

Fleischaker (1990, 1991) proposes that the concept of autopoiesis, together with notions of a *minimal cell*, can provide a sound theoretical framework to tackle these questions within the second tradition mentioned above. Autopoiesis clearly does aim to provide a specific and operationally useful definition of life, although Fleischaker (1988) argues that the concept of autopoiesis does need some modification. This modification would restrict "living" systems to autopoietic systems in the *physical domain* rather than allow the possibility of nonphysical living systems, a possibility which (as mentioned above) is left open by the formal definition of autopoiesis. This will be discussed in Section 3.3.2.

Given autopoiesis (or a modified version) as a definition of *life*, the next step in theorizing about the *origin* of life is to consider how an elementary autopoietic system might have formed. Note that autopoiesis is all or nothing. A self-producing system either exists and produces itself or it does not—there can be no halfway stage. This leads to the idea of a theoretical "minimal" cell which could plausibly emerge, given the early conditions on earth. In fact, Fleischaker (1988) considers three different characterizations of minimal cells: a minimal cell representative of the evolved life forms that we know today; a minimal cell that would characterize the origin of life on earth; and an abstract minimal cell that would characterize both terrestrial and nonterrestrial life regardless of its constituents.

About the last, little can be put forward beyond the six-point autopoietic characteristics in the physical space; to be more specific would constrain the possibilities unnecessarily. On the other hand, we can be quite specific about a modern-day cell. Such a cell could be described as "a volume of cytoplasmic solvent capable of DNA-cycled, ATP-driven, and enzyme-mediated metabolism enclosed within a phospho-lipo-protein membrane capable of energy transduction" (Fleischaker, 1988, p. 45). This generalized specification can cover both prokaryotes (bacteria) and eukaryotes (algal, fungal, animal, and plant cells) even though there are important differences in their operation.

The most interesting minimal cell scenario concerns the origin of life. The first cell need be only a very basic cell without the later elaborations such as enzymes. Fleischaker (1990) suggests that such a cell must exhibit a number of operations (Fig. 2.4):

1. The cell must demonstrate the formation and maintenance of a boundary structure that creates a hospitable inner environment and allows selective permeability for incoming and outgoing molecules and ions. The lipid bilayer found in contemporary cells is a good possibility since the hydrophobic nature of lipid molecules leads them to form closed spheres in order to avoid contact with water. Lipid bilayers are also permeable in certain ways—for example, to flows of protons or sodium atoms—without the need for the complex enzymes prevalent in contemporary cells.

2. The cell must also demonstrate some form of active energy transduction to maintain it away from entropic chemical equilibrium.

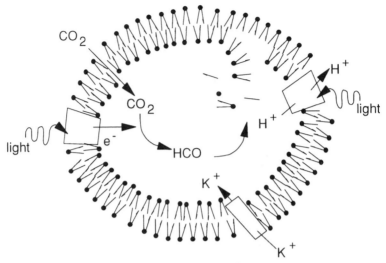

FIGURE 2.4. An Operational Minimal Cell (after Fleischaker, 1990, p. 135).

One possibility is an early form of photopigment system driven by light. Pigment molecules would become embedded in the membrane and act as proton pumps, leading to the concentration of a variety of raw materials in the cell.

3. The cell would also need to transport and transform material elements and use these in the production of the cell's components and its boundary. A possible start in this direction would be the import of carbon dioxide and the physico-chemical transformation of its carbon and oxygen through light-driven carbon fixation.

What is important is not the particular mechanisms for any of these general operations; but that whichever mechanisms are postulated, all operations need to be part of a continuous network to form a dynamic, self-producing whole.

2.4.2. Chemical Autopoiesis

Beyond theoretical constructs of minimal cells, it is also interesting to look at attempts to identify or create *chemical* systems based on autopoietic criteria, and to consider whether or not these are *living*. We shall look at three examples: autocatalytic processes, osmotic growth, and self-replicating micelles.

2.4.2.1. Autocatalytic Reactions

A catalyst is a molecular substance whose presence is necessary for the occurrence of a particular chemical reaction, or which speeds the reaction up, but which is not changed by the reaction. The complex productions of contemporary cells (as opposed to cells that may have existed at the origin of life) require many catalysts, and this is one of the main functions of the enzymes. An autocatalytic process is one in which the specific catalysts required are themselves produced as by-products of the reactions. The process thus self-catalyzes. An example is RNA itself which, in certain circumstances, can form a complex surface that acts like an enzyme in reaction with other RNA molecules (Alberts *et al.*, 1989). Kauffman (1993) has a detailed discussion within the context of complexity theory.

Although this process can be described as a self-referring interaction, the system does not qualify as autopoietic because it does not produce its own boundary components and thus cannot establish itself as an autonomous operational entity (Maturana and Varela, 1980, p. 94). Complex, interdependent chemical processes abound in nature, but they are not autopoietic unless they form self-bounded unities that embody the autopoietic organization.

2.4.2.2. Osmotic Growth

Zeleny and Hufford (1992a, 1992b) have suggested that a particular form of osmotic growth, studied by Leduc (1911), can be seen as autopoietic. The growth is a precipitation of inorganic salt that expands and forms a permeable osmotic boundary. This can be demonstrated by putting calcium chloride into a saturated solution of sodium phosphate. Interaction of the calcium and phosphate ions leads to the precipitation of calcium phosphate in a thin boundary layer. This layer then separates the phosphate from the calcium, water enters through the boundary by osmosis, and the increased internal pressure breaks the precipitated calcium phosphate. This break allows further contact between the internal calcium and the external phosphate, leading to further precipitation. Thus the precipitated layer grows (Fig. 2.5).

Zeleny and Hufford argue that this system fulfills the six autopoietic criteria:

1. It is a distinguishable entity because of its precipitate boundary.
2. It is analyzable into components such as the calcium phosphate boundary and the calcium chloride.
3. It follows mechanistic laws.

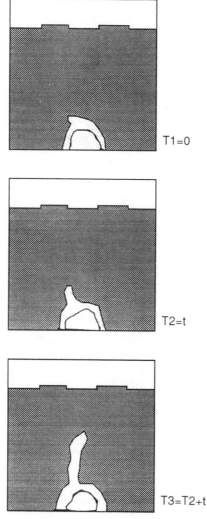

FIGURE 2.5. Osmotic Growth (after Zeleny and Hufford, 1992a, p. 152).

4. The boundary components (calcium phosphate) aggregate because of their preferred neighborhood relations.
5. The boundary components are formed by the interaction of internal and external components following osmosis through the membrane.
6. The components (calcium chloride) are not produced by the cell but are permanent constituent components in the production of other components (the precipitate).

This hypothesis does cause problems, as Leduc's system is clearly inorganic and not what would be called living. If it is accepted that the system does properly fulfill the criteria of autopoiesis, i.e., that it is an autopoietic system as currently defined, then either we must expand our concept of living or accept that autopoiesis is in need of redefinition to exclude such examples. In fact, it is debatable (Fleischaker, 1992a, 1992b) whether or not this osmotic growth does correctly fulfill the six criteria. It certainly meets the first three, but it is not clear that it is a dynamic network of processes of production.

As for the fourth criterion, the precipitate that forms the boundary is unlike a cell membrane. It is static and inactive, more like a stone wall than an active membrane. It is not formed through "preferential neighborhood interactions"; in fact, once formed, it does not interact at all. Considering the fifth criterion, the boundary components are not continuously produced by the internal processes of production. Rather, a split or rupture occurs and more boundary is precipitated at the split through the interaction of internal and external chemicals. It is only because of, and at, the rupture that new boundary is produced. Finally, as for the sixth criterion, there is only one internal component, calcium chloride, which is introduced artificially at the beginning, is not produced by the system, and eventually runs out.

Although the debate is not over, in my own view this osmotic system is not a dynamic autopoietic system. Rather, it is a particular form of static crystallization which happens to generate a spatial boundary and thus has a superficial structural resemblance to a living cell.

2.4.2.3. Self-Replicating Micelles

An approach with more potential, currently being researched by Bachmann and colleagues (Bachmann et al., 1990, 1991; Bachmann et al., 1991), was first proposed by Luisi (Luisi and Varela, 1989). It has been discussed by Maddox (1991) and Hadlington (1992). A micelle is a small droplet of an organic chemical such as alcohol stabilized in an aqueous solution by a boundary or "surfactant." A reverse micelle is a droplet of water similarly stabilized in an organic solvent. Chemical reactions occur within the micelle, producing more of the boundary surfactant. Eventually, this leads to the splitting of the micelle and the generation of a new one, a process of self-replication. Experiments have been carried out with both ordinary and reverse micelles and with an enzymatically driven system.

In the reverse micelle experiments (Fig. 2.6), the water droplets contain dissolved lithium hydroxide, one of the surfactants is sodium octanoate, and the other is 1-octanol, which is also a solvent. The other

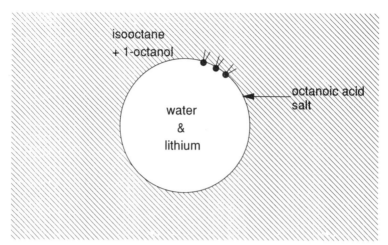

FIGURE 2.6. Reverse Micelle Reactions (after Bachman *et al.*, 1991, p. 266).

solvent is isooctane. The main reaction is one in which the components of the boundary are themselves produced at the boundary. Octyl octanoate is hydrolized using the lithium as a catalyst. This produces both the surfactants (sodium octanoate and 1-octanol). Since the lithium hydroxide is insoluble in the organic solvent, it remains within the water micelle, thus confining the reaction to the boundary layer. Once the system is initiated, large numbers of new micelles are produced, although the average size of the micelles decreases.

It is not clear that these systems could yet be called autopoietic. First, the raw materials (the water–lithium mixture or the enzyme catalyst) are not produced within the system. This limits the amount of replication which can occur; the system eventually stops. Even if these materials could be added on a regular basis, the system would still not be self-producing. Second, the single-layer surfactant does not allow transport of raw materials into the micelle. For this to happen, a double-layer boundary would be necessary, as exists in actual cell membranes. Moreover, the researchers themselves seem most interested in the fact that the micelles *reproduce themselves,* and seem to identify this as autopoietic. However, reproduction of the whole is quite secondary to the autopoietic process of *self-production of components* (see Sections 2.4.1 and 3.2.4). Nevertheless, this does represent an interesting step toward generating real autopoietic systems.

3

The Implications of Autopoiesis

Plus ça change, plus c'est la même chose
French proverb

The basic nature of living systems as autopoietic has now been described. They are characterized by a circular organization of production processes that continually produce and replace the components necessary to that organization. There are a number of important implications of this theory, which will now be sketched out. Some—for example, autopoietic organizations and the nature of information and cognition—will be expanded on in later chapters.

3.1.   Structure-Determined Systems and Organizational Closure

One of the main principles underlying the concept of autopoiesis is that of *structural determinism* as well as the related idea of *organizational closure*. These concepts, particularly the second, have led to considerable misunderstanding, and Maturana has tried to clarify this in recent publications (Maturana, 1991a; Krull *et al.*, 1989). I shall first briefly recap the difference between structure and organization. By *organization* Maturana refers to the relations between components that give a system its identity, that make it a member of a particular type. Thus, if the organization of a system changes, so does its identity. By *structure* Maturana means the actual components and relations between components that constitute a particular example of a type of system. The organization is realized through the structure, but it is the structure that can interact and change. So long as the structural changes maintain the organization, the system's identity remains.

3.1.1. Structure-Determined Systems

In considering change in a system, Maturana argues that all composite systems (i.e., those consisting of components) are structure-*determined*. He means by this that the actual changes that the system undergoes depend on the structure itself at a particular instant. Any change in a composite system must be structural change—i.e., it must be a change in the components or their relations—and, as such, must be determined by the properties of the components. Changes occur in response both to internal dynamics and to interactions with external systems, but even in external interactions the resulting change is determined internally; it is only *triggered* by the environment. This is a very important conclusion, for it means that there can be no "instructive interactions." That is, it is never the case that an environmental action (be it physical or communicational) can determine its own effect on a structure-determined system.

> In general then, everything that happens in a composite unity is a structural change, and every structural change occurs in a composite unity determined at every instant by its structure at that instant. . . . It follows from all this that composite unities are structure determined systems in the sense that everything is determined by their structure.
> Maturana (1987, p. 336)

Maturana talks of perturbations in the environment only triggering structural change or *compensation*. It is the structure that determines both what the compensation will be and even what in the environment can or cannot act as a trigger. In total, the structure at any point in time determines

1. all possible structural changes within the system that maintain the current organization, as well as those that do not, and
2. all possible states of the environment that could trigger changes of state and whether such changes would maintain or destroy the current organization.

Looking at some examples will show that this is reasonably straightforward, although we are accustomed to seeing things in a different way. The examples are biological, but the concept applies to all composite systems.

Living things are continually changing and developing, and these changes are determined by their own structure. Some changes, such as growth, leave the organization the same; other changes result in a new organization—for example, a caterpillar developing into a butterfly or an egg into a chicken—while others, such as death, lead to the loss of both the organization and the unity. Equally, what does or does not affect the organism and the nature of any effect is determined by its structure. Humans have receptors for light and color and so can be triggered by it, while bats can receive high-pitched sounds that humans cannot hear. Each organism has its own particular domain of interactions that can affect it and those which cannot. The effects are also structure-determined. Berries that are poisonous to humans are food for other animals; carbon dioxide is necessary for plants but inimical to humans, while oxygen is the opposite. In each case, the nature of the effect of a particular substance is determined not by the substance but by the organism. We humans often label things as poisonous and think that this quality is intrinsic to the substance when, in fact, it is not.

3.1.2. Organizational Closure

All composite systems are constituted by an organization and realized in a structure. But within this general class there are some systems that Maturana and Varela have termed *organizationally closed*, such as the nervous system, the immune system, or a social system (Krull *et al.*, 1989). Autopoietic systems are also organizationally closed, but they are a specific type in that they are also self-producing (see Fig. 3.1). This is discussed further in Section 6.4.

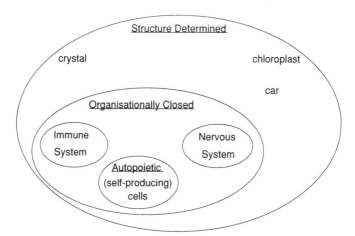

FIGURE 3.1. Types of Systems.

A system is organizationally closed if all its possible states of activity must always lead to or generate further activity within itself. In an autopoietic system, all activity must maintain autopoiesis or else the system will disintegrate. All processes are processes of self-production; the system's activity closes in on itself. Similarly, Maturana and Varela argue that the defining feature of the nervous system is that it is closed (see Chapter 5). All states of neuronal activity lead to further neuronal activity. All neurons both affect, and are affected by, others. Even the motor and sensor neurons are no exception to this; they do not "open" the nervous system to the environment. The motor neurons trigger sensor neurons through the activity they initiate, and sensor neurons are thereby also internally stimulated.

The particular nature of organizationally closed systems can be shown with a simple example. Consider a gas boiler and the pilot light used to start it. The gas heater itself is organizationally open; it takes in gas and puts out heat. However, the pilot light is closed (see Fig. 3.2). When it is on, the pilot light heats a thermocouple connected to the gas supply, which in turn fuels the pilot light. Clearly this is a self-dependent, all-or-nothing system. If the pilot light is not on, it does not heat the thermocouple, and no gas is supplied, so the pilot cannot be on.

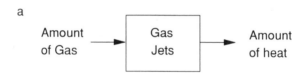

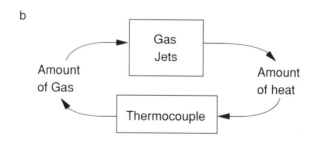

a) Gas Heater

b) Pilot Light

FIGURE 3.2. Open and Closed Systems.

This is why to light a gas boiler it is necessary to intervene from the outside, to break into the closed system and manually provide gas until the necessary temperature is reached. Once working, the system will carry on, enabling itself until some other external force intervenes. This is organizationally closed but is certainly not autopoietic because it does not produce a boundary nor any of its own components.

It can also be said that organizationally closed systems do not have inputs and outputs (Maturana and Varela, 1973)—or, at least, are not characterized in such terms (see discussion below). Autopoietic systems are organizationally closed because the product of their organization is that very organization itself. They do not primarily transform an input into an output except in the sense of transforming themselves into themselves. All the possible states that they can enter must conform to or maintain the autopoietic organization; otherwise they disintegrate. It may appear that the structure of an autopoietic system changes in relation to, or in response to, changes in its environment. But, for an observer to see such changes in the environment as an *input* and the structural change as an *output* is to mischaracterize the system as allopoietic, since the changes will, actually, have been devoted to maintaining autopoiesis. Such a description pertains only to the observer who can witness both the organism and its environment and relate the two.

The notions that autopoietic systems are organizationally closed and have no inputs and outputs have often been misinterpreted. They have been taken to mean that such systems are completely isolated and have no interactions with their environment. This is not at all the case. Such systems are organizationally closed but *interactively* open. They interact with their environment through their structure.

> I use the expression organizational closure to refer to systems whose organization is closed but whose structure is open in order to highlight the fact that I am talking of their organization and not of their structure.
> Krull *et al.*, (1989 p. 91)

Cells, for example, take in raw materials and energy and excrete waste products. Their structure is open to interactions; their organization is closed.

3.1.3.   Inputs/Outputs and Perturbations/Compensations

It is easy to become confused about inputs/outputs and perturbations/compensations in Maturana's and Varela's writings. Autopoietic systems can only be perturbed by their environment, yet they appear to have structural inputs and outputs. The problem is partially clarified in Maturana (1991a), where it turns out that there are really two different questions involving input and output. The first concerns the organization of the system (whether it is closed or open), and the second concerns the level of description of the observer (whether we are interested in characterizing the unity itself, as a composite entity, or whether we are interested in how the unity as a whole relates to some wider system).

First, all dynamic systems interact with their environments through their structure. In identifying a system as a composite entity, we are interested in the organization of the entity and its particular characteristics. Do we need to characterize the organization (and thus the entity) by inputs and outputs? If we are describing the autopoietic organization (and other closed organizations) then we do not. But if we are describing an organization that is not closed, then its inputs and outputs are necessary for a proper characterization. So, for example, organizationally, the cell is closed and has no inputs and outputs but the heart, as a blood pump, does.

The second question is whether we should refer to inputs and outputs or perturbations and compensations. This depends on whether we describe the entity in its own right or as part of a wider system. In the former case, as all systems are structure-determined, all interactions should be described as perturbations that lead to particular compensations. However, if the entity is part of a wider system, which produces repetitive pertubations, then, from that wider point of view, the interactions may be seen as inputs and outputs. For example (from Maturana 1991a), for a thermostat seen as a composite entity, a heat rise is a perturbation, but for a thermostat seen as part of a heating system, the heat rise is an output. Equally, a cell can be seen allopoietically as part of the liver, in which case it can be described in terms of inputs and outputs.

There is thus the organizational question of whether an entity is characterized by inputs/outputs and the structural question of whether the level of description requires inputs/outputs or perturbations/compensations.

3.1.4. Structural Coupling

Structural determinism implies that it is wrong to suggest that the environment determines or specifies the changes of state of the system. This is difficult to accept initially because it appears as though organisms are so well adapted to their environments that the environment must have led to appropriate changes in the organism. The answer lies in Maturana's important concept of *structural coupling* (Maturana 1978a, p. 35; 1980a, p. 70).

As we have seen, an autopoietic system is realized through a particular structure, and the changes that it can undergo are determined by that structure so long as autopoiesis is maintained. These changes may preserve the structure as it is, or they may radically alter it (think of an acorn developing into an oak). Where this is possible, the structure is said to be *plastic*. This plastic structure exists within an environment that perturbs it and can trigger changes. The environment *does not determine* the changes, but it can be said to select states from among those *made possible* at any instant by the system's structure. In an environment characterized by recurring states (and an actual autopoietic system will require, for example, a continual availability of energy), continued autopoiesis will lead to selection in the organism of a structure suitable for that environment. The organism becomes structurally coupled to its environment and, indeed, to other organisms within that environment. Structural coupling is a reformulation of the idea of adaption, but with the important proviso that *the environment does not specify the adaptive changes that will occur*. They either will occur, and thus maintain autopoiesis, or they will not, and the system will disintegrate:

> . . . if a composite unity is structurally plastic, then adaption as a process of structural coupling to the medium that selects its path of structural change is a necessary outcome.
> Maturana (1981, p. 29)

At first sight this sounds rather abstract, but examples will show how commonplace it is. The first example is that of a person interacting with a particular computer program. We can say that the computer and its software is itself an example of a structure-determined system that is interactively open but organizationally closed. The person can interact with the computer and can type in information and get appropriate responses. However, the computer is structure-determined since it is the structure of the program and that of the computer that determines what will or will not trigger it. Only pressing appropriate keys (or the like) will lead to appropriate responses, and those particular triggering mechanisms are determined entirely by the nature of the system. Even simple operations of a similar nature vary from one software package to another.

When beginning a new package, one has a feeling of insecurity, not knowing how to achieve what one wants, not knowing whether one has performed the right actions, pressed the right keys. Gradually, through use, this feeling disappears until eventually one reaches a state in which it is almost unnecessary to think about the actual operations; one merely needs to think of what is to be achieved. This state of being able to interact without thinking consciously of what to do is called by Heidegger (1962) a state of "being thrown." This process of becoming attuned is, in fact, the process of developing structural coupling.

The second example is the development of babies. In their first few months they are becoming structurally coupled to their physical environment. Their structures are developing in ways that reflect the interactions they have with their environment. Then, up to three years of age, they also become structurally coupled in the linguistic domain. This is a very important domain of activity for human beings—indeed, it is probably their primarily distinguishing characteristic from other animals. Maturana describes language (see Section 6.3) as a consensual domain, implying that the tokens we use in our language do not have meaning of themselves but depend on a consensus among the people involved in using the language. This, of course, requires structural coupling. Once we have learnt a language, we feel so comfortable and easy in using it that it appears as though the language and the words have inherent meaning in their own right. However, the above description reveals the true nature of the situation—namely, that communication is possible only to the extent that the systems involved are structurally coupled.

If structure-determined coupling is actually so obvious, why is it of importance? First, it shows that all interactions that we have as human beings, as autopoietic systems, are determined by our own structure. Things in our environment can be triggers for the nervous system only if the nervous system can react to them, and the reaction they get depends on the state of the nervous system. We cannot, therefore, have interactions with anybody or anything that are in some sense pure—they all are generated by our own nervous system.

Organisms become structurally coupled not only to their medium, but also to other organisms. The behaviors of one become triggers for the behaviors of the other through the selections of their individual structures. These interlocked triggering behaviors may have direct importance in themselves, as in the case of a threatening gesture and a corresponding flight, or they may be purely symbolic and essentially arbitrary, such as a particular form of greeting in a particular language. In the latter case, it does not matter what the actual behavior is, only that it has been implicitly agreed through structural coupling. This idea is of great importance in Maturana's cognitive theories, as it is the basis for his concept of a consensual domain, that is, a domain of behaviors (including, above all, language and description) which are both arbitrary and context dependent (see, for example Maturana, 1978a, p. 47).

Finally, Maturana calls the dynamic outcome of structural coupling for a particular system *ontogenic structural drift* (Maturana, 1987, p. 344). The analogy is with a boat, drifting uncontrolled in the sea, whose path is continuously determined by its structure and the effects of wind and waves. It is equally so for a particular structure-determined system. Such a system, in interaction with its environment conserves its organization through structural coupling. Its particular structural changes are triggered by occurrences in its environment and, just as the path of the boat is a determinate outcome of its history, so is the path of structural change of the system.

3.2.   Biological Implications

There is a worldview within biology, perhaps the dominant one, which places genetics and evolution at its core. The classic work of Dawkins (1978) views the gene and its survival and development through evolution as the centerpiece of life. Individual organisms and groups of organisms are of secondary importance. This approach tends also to employ a functionalist mode of explanation (Lambert and Hughes, 1988), which suggests that particular traits or components come about in order to fulfill a need posed by the environment. Maturana's and Varela's work, which can be seen as an example of a structuralist approach to biology, presents quite antithetical views. Life is a property of the individual, autonomous entity, such as the cell; reproduction and heredity are a secondary development of living organisms; and functionalist explanation is eschewed.

3.2.1.   Autonomy

Autopoietic systems are autonomous—they depend essentially only on themselves for their continued production, and physically they define themselves through the production of their own boundaries (Varela, 1976, 1977c, 1981b, 1984a; Tabary, 1991). This occurs independently of an observer whose description may or may not correspond to these boundaries. The interactions an autopoietic system can undergo are also determined by its autopoietic organization. Interactions that do not allow the continuance of autopoiesis lead to its disintegration. An autopoietic system also has individual identity since, so long as it follows a continuous process of autopoiesis, it maintains its organization despite significant change in its appearance (its structure).

Allopoietic systems, conversely, do not define their own organization but depend on an observer to determine their identity. They rely on other systems for their continued production, and the result of their activity is something other than themselves. This is not to say that autopoietic systems cannot be treated as allopoietic either by an observer or by other entities with which they interact. For example, autopoietic cells do play an allopoietic role within multicellular organisms, but this in no way diminishes their primary autopoiesis.

3.2.2.   *Autopoiesis as Nonfunctionalist*

The actual processes that occur in a living organism depend only on the immediate neighborhood interactions and reactions of the components involved and do not in any causal sense depend on a reference to, or representation of, or any supposed functions of, the system as a whole. Autopoiesis specifies certain necessary conditions and relations, and if these arise then an autopoietic unity is established. This is entirely a contingent matter, however, and there is nothing in the theory of autopoiesis to suggest that autopoiesis brings about or causes particular structures to arise. There is no need for functionalist explanations or teleonomic ideas such as purpose in the explanation of living things, although they may be useful in the descriptive language of an observer who sees the components, the unity, and its history of development:

> . . . if living systems are physical autopoietic machines, teleonomy becomes only an artifice of their description which does not reveal any feature of their organization, but which reveals the consistency in their operation within the domain of observation. Living systems, as physical autopoietic machines, are purposeless.
> Maturana and Varela (1980, p. 86)

This may seem contrary to what we observe, namely, the apparent fit or adaption of organisms to some independent environment. It is this which makes functionalist explanations attractive—the existence of treetops leads to the development of long necks in giraffes—but there never is such a causal relationship. Rather, Maturana developed the concept of *structural coupling* to explain the complementarity between organisms and their environments, as explained in Section 3.1.4.

3.2.3.   *Reproduction, Heredity, and Evolution*

Reproduction has generally been seen as a defining feature of living systems, but Maturana and Varela (e.g., Maturana, 1980a; Maturana and

Varela, 1987) show that reproduction (i.e., the production of another entity) can come about only after the formation of a unity and is, therefore, derivative from it. Moreover, it is only with reproduction that the concepts of *heredity* and *evolution*, and indeed of *species*, can have meaning. Therefore, these too are not the primary features of living systems but are secondary to the establishment of a single autonomous autopoietic entity.

The essence of reproduction is the production of another, distinguishable, entity of the same class as the first. This is, of course, quite distinct from autopoietic processes internal to the continual production of a single unity. Logically, therefore, reproduction requires the existence of an entity to be reproduced. However, this does not imply that reproductive capacity is a necessary characteristic of the living organization. There can be living organisms that are biologically incapable of reproduction, such as the mule (a cross between a donkey and a horse). Thus, the fact that the overwhelming majority of organisms can reproduce is not a defining feature of life. Rather, it reflects the simple logic that those that can reproduce outnumber, over time, those that cannot.

Once reproduction is possible and once there is a mechanism by which offspring both inherit characteristics of the parents yet can be different, then the whole process of heredity and evolution begins. How does the phenomenon of reproduction relate to the autopoiesis of living systems? It is certainly not necessary, but it is quite compatible with continued autopoiesis. Maturana and Varela (1987, p. 59ff) distinguish three ways in which further entities of a particular type may be produced: *copying, replication,* and *reproduction.*

Copying is a process whereby one particular entity is used as a basis for generating an imitation or likeness of it, for example, bacterial cloning or machine photocopying. In the latter, there are two different situations to be considered—that in which the copies always come from the same original and that in which a copy becomes an original for the next copy. In the first instance, there is no history or heredity from one copy to another—the copies are independent of one another although all are related to the original. In the second, however, there is a history of transformation. The copy will progressively change, becoming more and more different from the original until, in the end, it may become unrecognizable as a copy of the original, just as happens in the children's game of Chinese whispers. This is a simple example of heredity, or what Maturana calls *historical* or *structural drift.*

Replication is the process whereby a number of similar but independent entities are produced by a particular, assembly-line, process: for example, cars in a factory or proteins in the cell. Once produced, all the entities are similar to one another (although not identical), but essentially independent—what happens to one does not necessarily affect the others. This process produces no linked history except insofar as the production process itself may slowly change over time.

Reproduction is a process in which an entity splits itself in two in such a way that two new entities *of the same type as the original* result. This is clearly what the cell does in mitosis or cell division, but many other systems also reproduce themselves, for example, a river splitting into two, a flatworm cut in half, or a piece of chalk breaking. What is necessary is that the essential features of the system are either distributed evenly throughout it, as with chalk, or can be replicated and compartmentalized, as with the cell. In the case of the cell, it generates its own compartmentalization as part of its autopoietic process. The nuclear membrane breaks down, DNA and other components appear in both halves of the cell, and eventually it splits into two equally endowed entities. These are similar but not identical, each partaking in the historical process and beginning its own ontogeny.

From an autopoietic perspective, evolution and adaptation take on a different perspective. As outlined in Section 3.2.2, adaptation is not caused or determined by the environment; rather a process of mutual change, structural coupling, occurs, resulting over time in structural drift. Essentially, evolution is a process of *conservation of organization and adaptation*. All the nondestructive interactions of an autopoietic system must conserve its autopoiesis. However, with a plastic structure, changes and modifications are triggered, and where these are compatible with or indeed facilitate the process of autopoiesis, they are maintained and conserved. Equally, the environment itself is changing, partly in response to the organisms themselves, and there is thus a neverending reciprocal adjustment leading to the continued success of some groups and the dying out, through loss of autopoiesis, of others.

3.2.4. The Domain of Biology

In claiming to be the defining characterization of all living systems, autopoiesis has important implications for biology. Autopoiesis is defined as the organization of a whole (that is, a unity) that is realized through the properties and interactions of its components (structure). These two domains, however, are separate and irreducible. The particular components composing a living system undergo the interactions appropriate to their make-up and necessarily obey physico-chemical laws. They do not, however, specify the interactions appropriate to the unity when considered as a whole, since this depends on the autopoietic organization. The implications are that biology should properly be concerned only with autopoietic systems, not with their components:

> . . . biological phenomena are necessarily phenomena of relations between processes which satisfy the autopoiesis of the participant living systems. Accordingly, under no circumstances is a biological phenomenon defined by the properties of its component elements, but it is always defined and constituted by a concatenation of processes in relations subordinated to the autopoiesis of at least one living system.
> Maturana and Varela (1980, p. 113)

In other words, a phenomenon is defined as biological not because of its particular make-up (e.g., as an organic molecule or based on DNA), but because of its constitution as an autopoietic process. In particular, this means that the basic unit of biology is not the species, and certainly not the gene (Dawkins, 1978), but the lowest level of the individual organism that is self-producing.

3.3.   Other Possible Embodiments of Autopoiesis

The cell has been used as the main exemplar of autopoiesis, but there has been much debate about other embodiments of the autopoietic organization. First, we will examine higher-order biological systems, and then, more radically, nonphysical systems.

3.3.1.   Higher-Order Biological Systems

Having seen how cells embody the autopoietic organization, it is natural to question whether more complex organisms such as animals are also autopoietic, at a higher level. At first sight, it would seem that these multicellular organisms are bounded and produce their own constituents, namely various types of cells, and thus are autopoietic. However, this subject is more complex than might at first appear, and Maturana and Varela themselves do not give a definite or even consistent answer.

First, let us consider the way that simple autopoietic systems such as cells can become coupled with one another. As we have seen, an autopoietic system becomes structurally coupled to its environment through recurrent interactions. It can similarly become coupled to another unity that is in its environment. Indeed, for the first system, the second one is just part of its environment (Margulis, 1993). In this way, two or more unities can become coupled with one another to the extent that they become mutually interdependent. The autopoiesis of the individual systems comes to depend on the continued autopoiesis of the others. In this way a composite unity comes about consisting of a number of component autopoietic systems locked together. Such a system has obvious evolutionary advantage in providing a more stable medium for its components, and it is therefore not surprising that multicellular systems predominate. Varela and Frenk (1987) discuss how such an analysis of structural coupling can generate a theory of biological shape.

Such a composite system consists of coupled autopoietic systems and provides a necessary medium for their continued autopoiesis. However, is it itself autopoietic as a unity? In Maturana and Varela (1973), the first point made is that this is not necessarily so:

> . . . if such a system [generated through the coupling of autopoietic unities—JM] is not defined by relations of production of components that generate these relations and define it as a unity in a given space, but by other relations, either between components or processes, it is not an autopoietic system. . . .
> Maturana and Varela (1973, p. 108)

It is certainly not autopoietic simply by virtue of its components being autopoietic. If, however, it can be shown that the composite unity does have its own autopoietic organization, with its autopoietic components playing allopoietic roles, then it is a *second-order* autopoietic system. They go on to say unequivocally:

> This has actually happened on earth with the evolution of the multicellular pattern of organization.
> Maturana and Varela (1973, p. 110)

However, in a later paper Maturana (1980a) seems not so sure. He suggests that many multicellular organisms are second-order autopoietic systems whose components are not, interestingly, the cells that constitute them, but "molecular entities" produced by the cells. He does not specify what these might be. However, other multicellular organisms may not be autopoietic but be merely "colonies," that is, aggregates of autopoietic systems.

Finally, in Maturana and Varela (1987, p. 87), there is a further retreat and a change in terminology. They now speak of *metacellulars* rather than *multicellulars,* as collections of coupled cells, and include within the definition organisms, colonies, and societies. Moreover, the meaning of "second-order autopoietic system" is changed. Now all metacellulars are second-order systems because they consist of first-order autopoietic systems rather than because they are autopoietic in their own right. The question now becomes, "are some metacellulars autopoietic unities? That is, are second-order autopoietic systems also first-order autopoietic systems?" (Maturana and Varela, 1987, p. 87). Unfortunately, no answer is given. They are unable to say what molecular processes might constitute metacellulars as autopoietic systems. They are willing to say, however (and this is one of Varela's main themes), that such a system is organizationally closed—that is, "a network of dynamic processes whose effects do not leave the network" (Maturana and Varela, 1987, p. 89).

Maturana and Varela go on to describe *third-order structural couplings,* that is, recurrent interactions leading to structural coupling among independent unities. Examples include social insects such as ants or bees, animals that form herds and packs, and humans forming societies. Phenomena arising through third-order coupling are defined as *social phenomena.* If the organisms involved have nervous systems, the behavioral domain can become very complex, leading ultimately, as we shall see in Part II, to language, self-consciousness, and the observer.

For myself, I do not see too great a problem in describing organisms such as humans as autopoietic in their own right. Each has a boundary, the skin, whose components are continually produced by the unity, and many other components, both autopoietic (e.g., particular cells) and nonautopoietic (e.g., the heart), which are also self-produced. If the network of self-production as a whole (rather than that of the individual cells) is interrupted, then the organism dies. It seems to me appropriate to restrict the term "second-order" to such systems, which are autopoietic in their own right. What seems much more problematic, and will be discussed extensively later, is the nature of social phenomena such as organizations and societies and the question of whether these can be autopoietic.

3.3.2.   *Nonphysical Systems*

The original definition of autopoiesis (Section 2.2) specified self-production but did not specify what was to be produced. This leaves open the possibility of autopoietic systems in the nonphysical domain: for example, computer-generated autopoietic models (Section 4.4), human organizations and societies (Chapter 8), and abstract systems such as law (Chapter 9) or systems of ideas. There are two main questions here—can there be nonphysical autopoietic systems? And, if so, are such systems living?

The first question is still open to debate. Certainly Maturana and Varela believe that their computer model does embody an autopoietic organization, and so it should be counted as autopoietic. More contentiously, societies and law are sometimes claimed to be autopoietic. So let us assume that there might be nonphysical autopoietic systems and consider the second question—whether they should be called living. If we accept the basic theory, then it is certainly true that autopoiesis is a necessary and sufficient condition for a (physical) living system. Thus all (physical) living systems must be autopoietic. But this does not imply that all autopoietic systems must be living. There are three possible positions (Fig. 3.3):

1. that "autopoietic" is identical with "living" and that therefore all autopoietic systems are also living systems;
2. that living systems are a particular subset of autopoietic systems, i.e., living systems are autopoietic systems embodied in physical space, and
3. that we should restrict autopoiesis to physical living systems, i.e., autopoiesis exists only in the physical domain.

The first position implies that we have to regard as "living" nonphysical systems such as the computer program and societies, something which seems to go against our idea of "living." The second implies that autopoiesis is not identical with life. This seems unfortunate as the concept of autopoiesis has been created precisely in order to explain life, and, since it generates the phenomena we observe as living, it would seem a weakness if some autopoietic systems were not living. The third

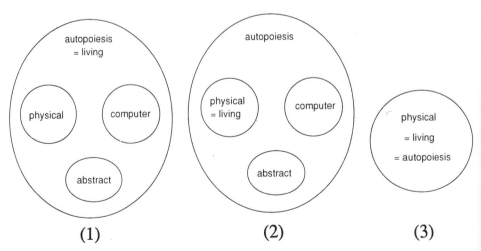

FIGURE 3.3. Possible Relations between Autopoietic and Living Systems.

seems unduly restrictive in placing an arbitrary limit on the nature of self-producing systems.

Maturana and Varela themselves have teetered between the first two positions. In earlier papers they argued for the second possibility:

> We also maintain that an autopoietic system in physical space . . . is a living system, and, therefore, that a living system is an autopoietic system in physical space. . . . There is no restriction on the space in which an autopoietic system may exist. The physical space in which living systems exist is only one of many.
> Maturana (1981, pp. 22–23)

However, Maturana (1991a) later explained that they had done this so as to avoid confusion, and that really any autopoietic system is living:

> I realized that it was necessary to make the molecularity of living systems explicit in order to avoid confusions. A computer model of an autopoietic system does not take place in a molecular space . . . and this is why we did not claim to have a living system in the computer. . . . Yet . . . since it is autopoiesis which defines them as the kind of system that they are, it could have been proper to claim that all autopoietic systems, regardless of the space in which they occur, are living systems. Indeed, this is what we at first wanted to do, but we thought that we would avoid much confusion if we were more conservative.
> Maturana (1991a, p. 376)

Fleischaker (1988) argues for the third position, suggesting that the definition of autopoiesis should be changed to specify that autopoietic systems can be realized only within the physical domain, based on processes of energy creation and use. In this, there is some support from Varela who, in later work (e.g., 1981a, p. 38), argued that notions of production did not seem appropriate for, e.g., social systems, and that autopoiesis should be confined to cells and animals.

This important debate mirrors one within a newly developed field known as "artificial life" (Langton, 1988; Langton *et al.*, 1991; Emmeche, 1992; Varela and Bourgine, 1992), which is also discussed in Section 4.4. This field is akin to the domain of artificial intelligence, which is premised on the idea that "intelligent behavior" can be separated from a particular medium, such as the brain, and generated on a computer. Artificial life makes a similar claim, that the logical form of life can be separated from its particular material embodiment, such as cells, and generated on a computer. Clearly, autopoiesis lies at the heart of this debate. For myself, I would tend toward the first, strong, position, that all autopoietic systems are living.

3.4.   The Role of Information and Representation

The ideas of a closed, structure-determined system and a consensual domain of essentially arbitrary behaviors have major implications for current beliefs and theories about the role that information and representations play in living systems and their thought processes. They challenge a number of current notions. First, for example, it is currently held that DNA and the genes code or contain or transmit information about the structure of their parent organism (the genetic code):

> These experiments, and other related ones, have finally brought us to a clear understanding of the nature of the unit of heredity. Like the dots and dashes of Morse code, the sequence of nucleotides in DNA is a code. The sequence provides the information that specifies the identity and order of amino acids in a protein. The sequence of nucleotides that encodes this information is called a gene.
> Raven and Johnson (1991, p. 305)

Second, it is currently held that the messages and communications between organisms are, in themselves, instructive: that is, that the messages contain sufficient information to determine an appropriate reaction on the part of the receiver. Third, a major plank of cognitive science, particularly as embodied in artificial intelligence (see Chapter 11 and Bobrow and Collins, 1975; Norman, 1981), is that our minds work by creating and then manipulating objective representations of the environment and the tasks to be performed within it. Cognition is seen as a process of symbol manipulation and information processing.

All of these ideas are so well established that they seem almost self-evidently true, yet autopoiesis suggests that they are all mistaken in the same fundamental way: they confuse the descriptions of an observer with the actual operations of an autopoietic system and ignore its structure-determined nature. As has been described above, autopoietic systems behave purely through their particular structure and the neighborhood interactions of their components at each point in time. Concepts such as "information" and "representation" pertain only to descriptions made by observers who can see both the internal interactions of a composite unity and the behavior of the whole in a particular environment and who can relate the two. So the idea that DNA contains or transmits information, or that the brain processes formal representations or symbols, must be purely metaphorical and does not describe how such systems actually operate in themselves. Also, since the environment can only trigger particular structural states, and not determine them, a successful message or communication must presuppose some correspondence in the domain of interaction of the two organisms—i.e., that they are already structurally coupled through their medium.

If the activities of autopoietic systems (including humans) are determined by their structure at each point in time, it could be argued that this constraint denies the idea of free will and implies a deterministic philosophy. This is a deep point, but Maturana's and Varela's theories about the nervous system and cognition are important here. They argue that the nervous system itself is organizationally closed and structure-determined—that is, that states of neuronal activity always stem from, and lead to, other states of neuronal activity in a self-referring manner. However, through the development of language, the possibility of making descriptions emerges. Such descriptions are associated with states of neuronal activity with which the organism can interact in a recursive manner, forming descriptions of descriptions, and so on. In short, this process generates the observer and, if the description can be a description of the self, the self-observer:

> The linguistic domain, the observer, and self-consciousness are each possible because they result as different domains of interactions of the nervous system with its own states in circumstances in which these states represent different modalities of interactions of the organism.
> Maturana and Varela (1980, p. 29)

Through this self-reflexive self-description lies the possibility of escape from predetermination.

3.5.   Epistemological Implications

Maturana and Varela are always aware of the epistemological implications of their ideas. These will be dealt with more thoroughly in Chapter 7, but a brief introduction to their ideas is in order here. As has been seen, they strongly maintain the distinction between the actual operational domain of an organism and the domain of descriptions of an observer: "Everything said is said by an observer, to another observer, who can be himself" (Maturana, 1975b).

Furthermore, Maturana's and Varela's cognitive theories show that the domain of description is inevitably relative to the describer. The observer can generate descriptions of her interactions, but they are embodied in states of relative neuronal activity and so must be subordinate to the organization and structure of the nervous system and ultimately to the autopoiesis of the organism. This means that we, as observers, can never escape from the domain of descriptions and thus have access to an absolute, objective reality. Rather, independent events may trigger a response or description, but the neuronal representation they lead to will always be structure-determined and thus inevitably relative to the observer.

Descriptions are not, however, completely individual:

> . . . the descriptions that the observer makes are made in the domain of consensual, observable behavior (linguistic domain), developed through a history of successful orienting interactions between two or more organisms.
> Maturana (1974, p. 468)

People develop similar cognitive structures because they undergo similar experiences in developing within a culture or society and because of the structural coupling that occurs among them within the consensual domain of language. This position, which Maturana terms *subject-dependent cognition,* is an example of a constructivist philosophy (Segal, 1986).

3.6.   Conclusions

This chapter has covered the main implications of the theory of autopoiesis in the physical domain. A number of these implications are quite radical in suggesting that traditional views of biology and of information and representation may be seriously misleading. Maturana's and Varela's further theoretical work on the nature of cognition and the nervous system and its implications will be explored in Chapters 5, 6 and 7, but in Chapter 4 we will take a slight detour to look at computer models of physical autopoiesis and their relationship to various formalisms such as Spencer Brown's "Laws of Form."

Mathematics and Models for Autopoiesis

4.1.   Introduction

Having considered the basic theory of autopoiesis, it is of interest to look at other forms of representation that have been developed for auto-poiesis—mathematics and computer models. In mathematics, Varela (1975) has extended Spencer Brown's *Laws of Form* (1972) into a *calculus for self-reference* and has then developed, with Goguen (Goguen and Varela, 1979), representations based on lattices, networks, and trees. *The Laws of Form* is itself a very unusual and contentious work, so Section 4.2 will try to provide some understanding of it before discussing Varela's extension. His later work with Goguen will be discussed in Section 4.3. Jumarie (1987) and Fontana and Buss (1994) have also developed some mathematics relevant to autopoiesis. There are interesting similarities with the work of Peirce on entitative and existential graphs (Roberts, 1973). Spencer Brown references Peirce's work in his book.

As for models, a computer model (a cellular automaton) was developed by Maturana and Varela to demonstrate autopoiesis, and this will be explored in Section 4.3.

4.2.   Laws of Form and the Calculus for Self-Reference

4.2.1.   Laws of Form—The Calculus of Indications

In this section I shall not, primarily, be providing a detailed exposition and explanation of the laws of form (even if I could) but shall attempt to outline and evaluate their meaning and significance—that is, what has been claimed for them and to what extent such claims are justified.

Spencer Brown's primary aim was to uncover what lies underneath logic and, in particular, Boolean algebra. The latter was created to provide a mathematical analysis of logic or, more precisely, Aristotelian syl- logistic logic, although it can be interpreted in different ways, as for example for set theory, sententional logic, or the logic of unanalyzed propositions. It will be useful to explain what is meant by logic because doing so will shed light on Spencer Brown's aims as well as the title of his work.

Logic can be defined as *the science of abstract form* (Lee, 1961, p. 13), where form is taken to be organization, pattern, structure, relationship. Form, as such, cannot be shown separately from a particular content but it can be studied, in the abstract, through an appropriate symbolism. "When the same form (organization of parts) can be abstracted from different concrete situations it is general. When we study it in its gen- erality, . . . then we have logic" (Lee, 1961). Thus logic can be seen as the study of relationship or order in general, abstracted from any partic- ular content. Boolean algebra was designed to fit logic, but Spencer Brown saw logic as separable from mathematics and one of his principle aims was "to separate what are known as algebras of logic from the subject of logic and to re-align them with mathematics" (Spencer Brown, 1972, p. xi). Spencer Brown's view of the importance of mathematics will be taken up at a later stage, but for the moment this aim can be realized by "a treatment of the form in which our way of talking about our or- dinary living experience can be seen to be cradled. It is the laws of this *form*, rather than those of logic, that I have attempted to record" (Spen- cer Brown, 1972, p. xx).

His approach began with the realization that Boolean algebra, as the name implies, is purely an algebra and that no one has ever studied the arithmetic on which this algebra is based. Briefly, the distinction between these is that an arithmetic uses constants whose values are known (e.g., 2, 5), whereas an algebra is concerned only with those properties of an arithmetic which hold irrespective of particular values [e.g., $a^2 - b^2 = (a - b)(a + b)$]. He therefore set about trying to discover the primary, nonnumerical arithmetic for Boolean algebra.

Logic and Boolean algebra deal with the form of linguistic state- ments. To find an arithmetic therefore means going beneath the level of language to uncover that on which language itself rests. Spencer Brown, like Maturana, sees language as essentially practical, not purely descrip- tive, and takes the most primitive activity as that of *indication* or distinc- tion. To distinguish *something* is the most basic linguistic act we can perform. Before counting things we must be able to distinguish between them, and before distinguishing several different things we must be able to distinguish something. This is the foundation of all language: to be able to create from nothing (the void) one thing, or state, or space that is distinct. The laws of form are concerned with the consequences of this most primitive act—the act of drawing a distinction. Spencer Brown prefaces his book with a quotation from Lao Tzu: *The nameless is the beginning of heaven and earth.*

What then is the nature of a distinction?

Distinction is perfect continence.

Spencer Brown (1972, p. 1)

This spare but very precise definition is characteristic of the flavor of the rest of the book. *Continence* is derived from the Latin *continere*, to contain, and the definition is saying that a distinction, i.e., the drawing of a boundary, perfectly separates that which is on one side from that which is on the other. This can be illustrated by drawing a circle on a piece of paper:

Out of the void we draw a distinction, and we can then separate that which is to be distinguished from everything else, and once it can be distinguished it can be indicated or identified. A distinction can be drawn only if there is some reason (intention or motive) for doing so, and there must therefore be some difference in value (to the person making the distinction) between what is distinguished and what it is distinguished from. We can give a name to what is distinguished, which then indicates the value. Saying or calling the name identifies the value and implies the distinction. Thus the act of indication, at this almost prelinguistic level, combines naming, acting, and valuing all in one. It is reminiscent of the protolanguage of Wittgenstein (1958, p. 3), where a builder shouts "slab" to his laborer and the laborer brings one. The shout distinguishes the slab from other things, values it, and generates an action.

As a direct consequence of this definition, Spencer Brown claims that two axioms can be stated—the law of calling and the law of crossing.

Axiom 1: The Law of Calling

The value of a call made again is the value of the call.

Axiom 2: The Law of Crossing

The value of a crossing made again is not the value of the crossing.

For Spencer Brown these capture the essence of a distinction—the difference between crossing and not crossing the boundary. The first axiom says that to draw a distinction and then to draw the same distinction again adds nothing new. To distinguish a circle and then distinguish a circle again leaves us with a circle. Thus, to re-call is to call. The second axiom involves us in crossing the boundary, in indicating the value by entering into the distinction. Now, if we draw another distinction, from within, we cross the boundary again and end up where we started, with no indication of anything. To distinguish a circle and then, from within, distinguish again must leave us with not-circle. Thus, to re-cross is to not cross.

One of the difficulties of grasping these ideas is that they are at a prelinguistic level, yet we inhabit language; therefore, as another example, consider a baby not yet able to talk. The baby cries (calls) for its mother's breast. This draws a distinction and values the contents. The distinction, the indication, and the value are one in the cry. The baby cries again and again, each cry drawing the same distinction and recalling the same call. This is in accordance with Axiom 1. Then the mother arrives, the baby goes to the breast, and the cries stop. The baby has entered into the distinction (mother's breast) and drawn another distinction (no longer need to cry for the breast), and peace returns. This is in accordance with Axiom 2.

These two rather unintuitive axioms form the basis of the whole calculus of indications. They are represented symbolically as initials of the arithmetic:

Initial 1: Number $\lceil\ \rceil$ = $\rceil$ Condensation

Initial 2: Order $\rceil\!|$ =   Cancellation

The mark $\rceil$ and the space   are the constants in Spencer Brown's arithmetic. $\rceil$ represents one of the distinguished states (the marked state) while   represents the other (the unmarked state).

Figure 4.1 is another illustration of the two axioms. In (a) we can see and mark a circle as a distinction. If we now put another mark within the distinction, as in (b), then we lose and gain nothing. We can still indicate the marked state, but there is still only one marked state. The second mark is redundant. This is in accordance with Axiom 1. Now, however, if we place another mark outside the boundary, as in (c) we can no longer distinguish the marked state; both states are marked. We thus lose the distinction. This is in accordance with Axiom 2.

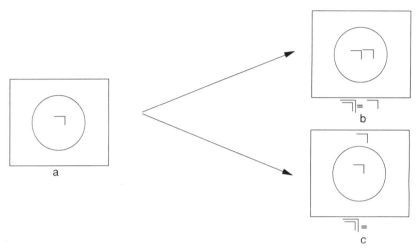

FIGURE 4.1. An Illustration of the Two Axioms.

Spencer Brown then shows that these two axioms alone are sufficient to generate a coherent and consistent arithmetic consisting of
expressions such as

$$\overline{\overline{\overline{\lnot}\,\overline{\lnot}}}\qquad\text{and}\qquad \overline{\overline{\overline{\lnot}\,\overline{\lnot}}}\;\lnot$$

which can always be simplified to either $\lnot$ or by applying the
two axioms.

He then goes on to demonstrate an algebra, based on this arithmetic, using the two algebraic primaries:

Initial 1: Position

$$\overline{\overline{p}\rvert\,p\rvert}\;=$$

Initial 2: Transposition

$$\overline{\overline{pr}\rvert\;\;\overline{qr}\rvert\rvert}\;=\;\overline{\overline{p}\rvert\,\overline{q}\rvert\rvert}\;r,$$

where p, q, r stand for any arithmetic expression.

Let us look at Initial 1. This says that whatever the value of p ($\lnot$
or), the expression always indicates the unmarked state. This is
like the law of the excluded middle. The inner part, $\overline{p}\rvert\;p$, is effectively
"p or not p". This must always be , since

$$\text{If } p = \lnot,\quad \overline{p}\rvert\,p = \overline{\lnot}\rvert\,\lnot\quad = \lnot$$
$$\text{If } p = \quad,\quad \overline{p}\rvert\,p = \lnot\quad = \lnot$$

The outer mark then negates this, so that nothing is "not (p or not p)." In other words, everything is either p or not p.

Having built an arithmetic and an algebra, it is necessary to show that it can serve as a basis for Boolean algebra. All Boolean algebras rest on a group of assumptions, or a postulate set, which are stated without proof. All the theorems in Boolean algebra can then be rigorously deduced from this postulate set (or, rather, these sets, as there are various versions). A postulate set consists of three elements:

   a. A set of undefined terms and relations between these terms.
   b. A set of terms and relations defined through (a).
   c. A set of postulates, which are very general statements about the "ways in which it will be legitimate in the system for the ideas and relations to go together" (Lee 1961, p. 258). These also are assumed and not defined.

There are numerous postulate sets for Boolean algebra (e.g., Huntington, 1904; Lewis and Langford, 1959; Sheffer, 1913), all of which are essentially equivalent and none of which has previously been proved. Spencer Brown takes Sheffer's set and shows that each postulate can be proved as a theorem in his algebra. Thus the whole of Boolean algebra (and its tremendously important applications in set theory and logic) can be shown to follow from the two axioms above and these, in turn, are a direct consequence of the primitive act of drawing a distinction. Spencer Brown also shows (as does Banaschewski, 1977) how the calculus of indications can be interpreted for logic, in particular the propositional calculus.

This certainly appears to be a very great achievement, not just because a simpler propositional set has been found, but because the axioms can be seen to be related directly to our common experience, in contrast to previous postulate sets, whose chief characteristic has been

> an almost total lack of any spontaneous appearance of truth. . . . The initial equations can be seen to represent two very simple laws of indication which, whatever our views on the nature of their self-evidence, at least recommend themselves to the findings of common sense.
> Spencer Brown (1972, p. xii).

Before moving on to Varela's extensions, I would like to mention briefly some of the possible applications of the calculus of indications. For example, it has been suggested that the calculus of indications can be interpreted and extended to elucidate other systems of philosophical logic. Orchard (1975) points to some possibilities. First, there are developments in propositional logic toward a sentential calculus with identity (Bloom and Susko, 1972). This takes the analysis of propositions beyond the point at which their truth value is discovered (when all true statements are logically equivalent) to recognize that the propositions might not be equivalent. Second, Kosok (1966) has attempted to formalize Hegel's dialectical logic in order to provide an account of the process of reflection, and this work too is based on the acts of indication and drawing a distinction. This is part of a more general area of study, that of many-valued (rather than merely two-valued) logics—see Rescher (1969) and Varela (1979b).

Moreover, and potentially this is the most significant claim, it has been suggested that the laws of form provide an important foundation for understanding human knowledge. The argument, with variations, runs that indication and distinction are essential elements in our perceptions and conceptions of the world and that the laws of form are therefore the laws of our description of the world and therefore of our knowledge. Spencer Brown thinks that mathematics is a very special subject in that mathematical forms represent and are derived from internal ways of thinking and that these have at least as much validity as knowledge of external reality. "The discipline of mathematics is seen to be a way . . . of revealing our internal knowledge of the structure of the world" (Spencer Brown, 1972, p. xiii) and later suggests that we have "a direct awareness of the mathematical form as an archetypal structure" (Spencer Brown, 1972, p. xvi). This must be so, he says, because the unprovable statements must vastly outnumber the provable statements, and yet a mathematician doesn't attempt to prove a statement unless he or she is quite certain that it can be proved even though he has not yet done so.

That this is so is important because Spencer Brown is attempting to explore this internal world. He believes that this is complementary to a study of the outer structure of reality, for "what we approach, in either case, from one side or the other, is the common boundary between them" (Spencer Brown, 1972, p. xvii), this boundary being the media through which we perceive the outside world. Thus the greatest significance of the laws of form lies in their explanation of this inner structure of knowledge in that they reveal the laws that must apply to our descriptions and understanding of the world, based, as they are, on distinction and indication. In this, Spencer Brown is pursuing a similar task to that of the phenomenologists such as Husserl (Section 7.5.1).

4.2.2.   Varela's Calculus for Self-Reference

Self-reference has been a problem in logic since the time of the Greeks, as it leads to paradoxical statements such as the problem of the Cretan liar. The problem arises because the proposition actually includes itself as a referent, that is, it refers to itself. These paradoxes tended to be ignored, but Russell and Whitehead (1927) found that they were actually crucial to the *Principia Mathematica*. Unfortunately, they could not provide any solution for them and therefore created a set of rules, the theory of logical types, to prevent such propositions from being considered.

Self-reference is also of central importance in the domain of living systems (Varela, 1977a), and particularly human beings, who have the capacity to build images of themselves, to be self-reflective. Varela was motivated to undertake his study because of his work on autopoiesis, in which epistemology must be self-reflexive since it is, itself, part of the area that is its subject, that is, knowledge.

The calculus for self-reference (CSR) has its genesis at a fairly advanced point in the calculus of indications (CI) where forms are allowed to enter their own expression (re-entering forms). That is where the value of an expression cannot be determined immediately because it is included within that expression, e.g.,

$$f = \overline{f}\rceil \quad \text{or} \quad f = \overline{\overline{f}\rceil}\rceil$$

The similarity to self-reference is immediately apparent; the form is self-referring.

The second of these two examples is unproblematic since $\overline{\overline{f}\rceil}\rceil$ is essentially a double negation and so is equivalent to f. However, the first is contradictory, since f must be equivalent to its own negation. Spencer Brown had to bring in the concept of time to overcome these problems by making f at one time depend on its own value at a previous time. This approach has been developed by Gunji and Nakamura (1991) and Gunji and Kon-no (1991). Varela (1975), however, introduced a new state (apart from the marked and unmarked states) which is autonomous and self-indicatory, i.e., it cannot be brought into being by the observer. The new state is now marked by the mark ⌐⌐ .

Varela reinterprets the whole calculus of indications employing this new state as a third arithmetic value. The initial axioms now become:

Initial 1: Dominance   ⌐⌐ = ⌐

                            ⌐⌐ = ⌐

Initial 2: Order   ⌐| =

Initial 3: Constancy   ⅂| = ⅂

Initial 4: Number   ⅂⅂ = ⅂

Varela then shows that the whole system of arithmetic and algebra remains coherent and consistent with this third value. It should be noted, however, that the system now implies a three-valued logic rather than the traditional two-valued logic.

The self-referential examples from the calculus of indications become

$$ f = \quad \rfloor \quad \text{and} \qquad f = \quad \rfloor| \quad = \quad \rfloor $$

At first sight this seems an important development and it has been used by Kauffman (1978) and Schwartz (1981) but, in a later paper, Varela and Goguen (1978) say "I do not now think that this represents a very satisfactory solution to the handling of re-entering forms," and list three reasons:

1. All infinitely re-entering forms are expressed by the same value (i.e., $\rfloor$), and we cannot therefore study the differences between those forms.
2. Certain forms in the CI no longer hold in the CSR and yet have great intuitive meaning, e.g., $\overline{\overline{p}|\ p}| = $, which expressed the law of the excluded middle—everything must be p or not p.
3. In trying to produce a corresponding three-valued logic, certain difficulties made the results unsatisfactory.

For these reasons, Varela explored other formalisms, which are discussed in Section 4.3. Turney (1986) has also proposed another interpretation of Spencer-Brown's re-entering forms.

4.2.3.   Evaluation of the Laws of Form and CSR

It must be said to begin with that, despite their apparent novelty and depth, the laws of form have made very little impact within logic and mathematics. This, in itself, does not refute the arguments—autopoiesis has made little headway in traditional biology—but there are few successful applications or developments to point to, and there are also a number of potential criticisms.

Both Cull and Frank (1979) and Kohout and Pinkava (1980) have made strong criticisms of the laws of form and Varela's extension to the CSR. The main ones are that there is no new original work, simply a reinvention of Boolean algebra in an obscure notation; that the notation itself is suspect; that the laws of form do not really solve Russell's paradox, as has been claimed; and that where the laws might be applied there are better methods. Kauffman and Solzman (1981) replied, rebutting some of these criticisms.

The main argument concerns the notation. It is claimed that Spencer Brown has two values (the marked and unmarked states) but only one symbol, ⌐ . The unmarked state is symbolized by the lack of a symbol—the blank page. It is thus not clear where it (the blank symbol) is or how it is manipulated. This both makes the notation look unusual and allows Spencer Brown to make some illegitimate moves. Moreover, the one symbol ⌐ is used both as a value (a constant) and as an operator. If this notation is replaced with conventional notation (e.g., ⌐ = 1,   = 0) then the primary axioms are seen to be basic Boolean operations such as disjunction or Boolean algebraic identities such as the excluded middle.

The reply to this is that the notation may be unusual, but this does not make it illegitimate. It is a notation based on geometry—the geometry of the plane and the rectangle—so it does follow rules and is consistent. It does make implicit assumptions about how it is to be read, but so does conventional notation (e.g., from left to right along a line). The mark does play a dual role, but it is part of the theory that a distinction both names and values, and the same happens in conventional notation (e.g., in 3 × 4 the 3 is both constant and operator—"three times . . .").

Cull and Frank (1979, p. 202) claim that at least two symbols are required to convey information, yet this is at the heart of Spencer Brown's work; drawing a distinction does create two states, but it is necessary to name only one of them, the other being distinguished by virtue of not being named. As for Russell's paradox, it is accepted that the CI does not actually solve it, but it does offer interesting ways of expressing self-reference.

My own main criticism is the fundamental idea of a "perfect" distinction. It was argued above (in Section 4.2.1) that as logic and knowledge rest on basic distinctions, if the laws of form follow from the act of making a distinction, then they could be seen as shaping logic and knowledge. However, I have my doubts as to whether this is really correct. Maybe it is the other way around. We value logic and therefore wish to prove its postulates; the postulates can be shown to depend on the two axioms, which are therefore necessary; the axioms depend on a perfect distinction; therefore we must assume the existence of perfect distinctions in order to validate logic.

But this does not prove that there are, or that we use, such things as perfect distinctions, and I myself am very doubtful about this. If we examine the physical world, it seems clear that there is a valid distinction between the table and not-table, although even here the distinction may not be so clear at the atomic or subatomic level. When we come to the social world and such concepts as good/bad, dominant/submissive, competitive/cooperative, can we really say that each of these distinctions is crystal clear and that people can be assigned to the different categories without any problem? The apparent usefulness of fuzzy set theory (in which the distinction is taken to exist but the boundary not to be perfect) seems to show that we cannot, and I think it would be of great interest to see how CI would have to be modified for it to lead to fuzzy set theory rather than classical set theory.

In summary, *The Laws of Form* is a very stimulating and rewarding work, but one that has not yet established its own validity either theoretically or empirically. It is certainly worthy of serious study.

4.3.   *Varela's Other Mathematical Representations*

In later papers other approaches are developed. Varela and Goguen (1978) consider the set of all the expressions possible in the CI in terms of an ordering based on the inclusion of one form within another. This includes infinite forms. Such a set is termed a *lattice*.

Goguen and Varela (1979) stress that there are two ways in which a distinction may be drawn and a system observed. We may choose to focus attention on the system *in its environment,* treating it as a unity with given properties, or we may choose to focus on the *internal constitution* of the system and view the environment as merely a background source of perturbations. These two views are, however, complementary, and in fact one can flip from one to the other as one moves through the levels of a hierarchy, from the cell to the organ to the organism to the population. At each level the distinction switches from the constitution to the unity.

They then show how there is a similar complementarity or duality in the mathematical forms of trees and networks. A network (mathematically, a directed graph) consists of nodes and links between them. A network can be used to represent the connections or relations among elements in a system. If we follow the permissible connections or links through the network, we trace the possible sequences of states of the system. These sequences or paths can be represented by a branching tree (mathematically, a reachable, loop-free pointed graph), the branches of which show the possible choices at each node. The tree thus unravels the dynamic possibilities of the network.

This is illustrated in Fig. 4.2, which shows a network consisting of four nodes (A–D) with six links (1–6) between them, and a complementary tree. The tree shows the possible paths starting from node A. Note that the tree is infinite.

They use this formalism as an illustration of the general complementarity described above and conclude that it is exemplified by the distinction between reductionism and holism, which should not be seen as conflicting alternatives, but as complementary descriptive viewpoints.

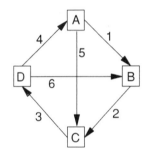

a   A Network

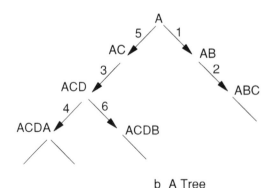

b   A Tree

FIGURE 4.2. Network Representations.

4.4.   *Computer Models of Autopoiesis*

Early on, Varela, Maturana, and Uribe (1974) developed a computer model to illustrate the autopoietic organization (see also Zeleny and Pierre, 1976; Varela, 1988, 1991b). As mentioned in Section 3.3.2, Maturana was willing to call this model "living." The model is a type of cellular automaton, that is, its universe consists of a (infinite) grid of cells, each of which represents a particular type of component. The model specifies rules for the generation and disintegration of these components, and also (in this case) for their movement around the grid. Varela (1988) has used another automaton to illustrate organizational closure.

In the autopoietic model the cells are initially occupied by "substrates" representing basic chemicals needed by a cell. These can move around the grid. There are also one or more "catalysts." A catalyst can convert substrate into "links"—the equivalent of components of the cell membrane—and can also move around. Links can bond with one another to form chains. Unbonded links can move, but bonded links cannot. Substrate can move through chains, but catalysts cannot. Links, bonded or not, may randomly disintegrate into substrate. The formal rules of the model are given in Table 4.1.

Dynamically, the catalyst starts to generate free links in its neighborhood. These move around and bond with one another to form a chain, which often closes on itself, creating a boundary around the catalyst, which cannot penetrate the links. The result is a "cell" which fulfills the autopoietic criteria. Substrate can pass through the bonds and so flows in and out of the cell. This means that more links can be formed inside. At some point, a link in the boundary disintegrates, leav-

TABLE 4.1
Rules of the Autopoietic Model

Composition	A catalyst and two substrates generate a link, leaving the catalyst unchanged. A hole also results.
Concatenation	A free link can join the end of a chain, two (bonding) chains may join together, or two free links may join.
Disintegration	A link, bonded or unbonded, disintegrates into two substrates in available wholes.
Other rules	The components move randomly across the surface subject to a precedence order (catalyst, free link, substrate), bonded links are fixed, and only substrate can move through links.
	Links disintegrate at random.

ing a space, but usually the free links generated inside fill the gap and recreate the boundary.

This is illustrated in Fig. 4.3, which shows successive stages of transformation in a computer run performed by the author. At time $t =$ 134, a cell has been formed around a catalyst. Substrate is inside. At $t = 135$, a link disintegrates, leaving two single-bonded links, a new free link is produced by the catalyst, and the catalyst moves into the hole thus formed. At $t = 136$, nothing changes, but at t = 137, the free link moves into the gap, recreating the boundary. The cell has reformed in a new shape. This illustrates the basic model but a great variety of examples may be generated by varying particular parameters such as the rate of distintegration. With more than one catalyst, several cells can be produced, and these may even join up to create multicellular entities.

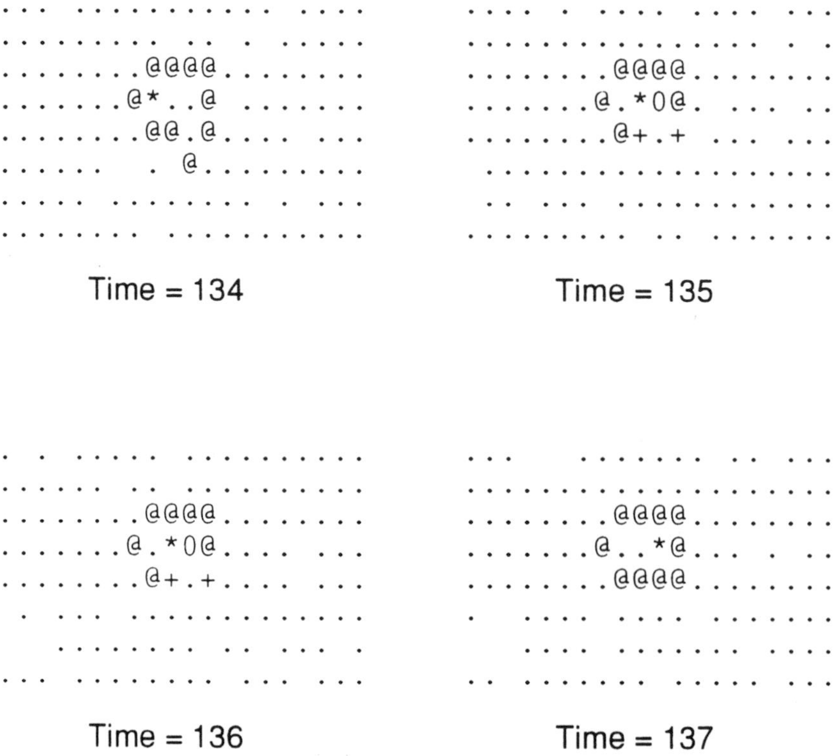

FIGURE 4.3. Four Successive Times in the "Life" of an Autopoietic Model. * = catalyst; + = single-bonded link; · = substrate; @ = double-bonded link; 0 = free link; = hole.

The model as a whole behaves as an autopoietic system. All the interactions are defined purely by the properties of the components and their local neighbors; there is no overall control, and yet the result is the establishment of a dynamic unity with a self-defined boundary producing its own components.

Although this work is quite old, there is currently a resurgence of interest through the development of a new area known as "artificial life" (AL) (Langton 1988; Langton *et al.*, 1991; Varela and Bourgine, 1992). This, in turn, is part of the exciting new field called complexity theory (Waldrop, 1992; Kauffman, 1993) which, in many ways, is a rediscovery of the basic systems ideas of emergence and order but with more effective mathematical and computing tools.

The idea of artificial life is analogous to artificial intelligence (AI), namely, that the logic or organization of "living" can be separated from the material base in which we normally experience it, and that, in particular, life, or lifelike behavior, can be realized computationally, i.e., in computer models (Emmeche, 1992). In other words, the aim is to create life as it could be rather than as it actually is. While this new venture is controversial (it produces debates similar to those of the strong version of AI), it is clear that autopoiesis is highly relevant—first, because of its separation of structure from organization and its recognition that the autopoietic organization can be realized in arbitrary structures, and second, because of the computer model we have just examined.

4.5. *Conclusions*

In this chapter we have looked at alternative representations of autopoiesis. Mathematically, we have seen attempts to model self-reference by using the laws of form, but it is not really clear at this point how valid or useful this has been. We have also seen a computer *model* of autopoiesis—or is it *really* an autopoietic, and therefore living, system? This debate, covered in Section 3.3.2, should now be more clear.

In the next section of the book, we move on to the cognitive aspects of autopoiesis, initially the nervous system.

II

Theories Of Cognition

The Organization of the Nervous System

5.1. Introduction

Part I of this book introduced the concept of autopoiesis in the physical domain and its implications. Equally important, and developed at the same time, are Maturana and Varela's work on the nature of cognition in autopoietic systems and their analysis of the organization and workings of the nervous system. Most of the original papers are by Maturana, so I will generally refer only to him.

Section 5.2 describes Maturana's theories about the nervous system and cognition. Section 5.3 explains the neurophysiology of the nervous system and some of its most important characteristics. The main consequences of this characterization of the nervous system, particularly the nature of perception and intelligence and the development of language as a subject-dependent domain, will be described in Chapter 6. The philosophical implications of Maturana's cognitive theories are considered in Chapter 7.

5.2. The Nervous System and Cognition

In general usage, *cognition* refers to the process of acquiring and using knowledge, and, as such, it is assumed to be limited to organisms with a (fairly advanced) nervous system. The nervous system itself is viewed as a system that has developed to collect knowledge about the environment, enabling an organism to better survive.

Maturana's theories question both of these beliefs (Maturana, 1970a, 1970b, 1975b, 1978a, 1978b, 1983, 1985, 1988b; Varela, 1971, 1977b, 1984b). He began his work in animal neurophysiology, publishing a number of well-known papers (Maturana *et al.*, 1960, 1968). This tied in with his biological interest in two seemingly unrelated questions: What is the nature of living organisms? And, what is the nature of perception and, more generally, of cognition and knowledge? A central breakthrough was seeing that the two questions are in fact linked. Perception and cognition occur through the operation of the nervous system, which is realized through the autopoiesis of the organism. As we have seen, an autopoietic system operates in a medium to which it is structurally coupled. Its survival depends on the continuance of certain recurrent interactions. For Maturana, this itself means that the organism has knowledge, even if only implicitly. The notion of cognition is extended to cover all the effective interactions that an organism has. The traditional dualism between knowing and acting is completely swept away—acting is knowing, and knowing is acting:

> A cognitive system is a system whose organization defines a domain of interactions in which it can act with relevance to the maintenance of itself, and the process of cognition is the actual (inductive) acting or behaving in this domain. Living systems are cognitive systems, and living as a process is a process of cognition. This statement is valid for all organisms, with and without a nervous system.
> Maturana and Varela (1980, p. 13)

I shall argue later that it is better to retain the more conventional use of *cognition*, but first I shall explicate the development and role of the nervous system and how it leads to the emergence of language and the observer.

5.3. The Nervous System

The nervous system is an evolutionary biological development, which increases the range of behavior that can be displayed by an organism—its requisite variety. It does not, in essence, change the nature of operation of an autopoietic system. We can see how nerve cells (neurons) have developed as specializations of ordinary cells. If we consider a single-celled organism such as the amoeba [Maturana and Varela (1987) and Von Foerster (1984) are very readable introductions], it displays behavior—for example, movement and ingestion. It has both a sensory and an effector surface—in fact they are both the same, its outer membrane. Chemical changes in areas of its immediate environment affect the elasticity of its membrane, in turn allowing its protoplasm to flow in a particular direction, thus leading either to movement or the surrounding of food. This eventually leads to the dying away of the environmental perturbation and the restoration of the status quo.

A neuron is like an ordinary cell except that it is specialized in two ways. First, it has developed very long extensions called *dendrites,* which connect to many other, often distant cells. This leads to a separation of the sensory from the effector sites of the cell and allows for the possibility of the transmission of perturbations. Second, it has developed a generalized response—electrical impulses (although neurons are still affected by chemical changes)—as opposed to the specific physicochemical sensitivity of other sensory surfaces. This has two vital consequences: the establishment of a universal medium (electrical activity) into which all the differing sensory/effector interactions can be translated and the development of internal neurons which connect only to other neurons, responding to this electrical activity. These *interneurons* are particularly important as they sever the direct relationship between sensor and effector and vastly expand the realm of possible behaviors of an organism. In humans these have grown to outnumber sensory/motor neurons by a factor of 100,000.

The other main physiological feature of the nervous system is the neuron's method of connection—the synapse. The synapse is the point of near contact between dendrites and other cells (neurons or ordinary cells). Any particular neuron/cell has thousands of synapses, each contributing a small amount to the cell's overall activity. A synapse is actually a very small gap across which chemicals called *neurotransmitters* can flow, triggering an electrical exchange. In effect, therefore, these are the sensory and motor surfaces of the neuron.

This organization has a number of consequences, which I will briefly outline. Some will be discussed more fully in later chapters.

5.3.1.   *Maintaining Internal Correlations*

What is it that the nervous system actually does? In the amoeba, a change in the sensory surface is triggered by the level of a chemical in the environment. This leads to motor changes and the movement of the organism through the environment. The process continues until the concentration is reduced and the balance between sensor and effector returns to the previous level. To the observer, the amoeba has moved or captured a prey. To the amoeba, state-determined structural changes have occurred restoring an internal balance or correlation between sensory and effector surfaces.

For Maturana, the nervous system functions in precisely the same way. It acts so as to maintain or restore internal correlations between sensory and effector surfaces. That it does so through an incredibly complex system of interacting neurons makes no difference to its fundamental operation. Touching a hot plate stimulates certain sensory neurons. These trigger motor neurons, leading to the contraction of a muscle. This in turn results in withdrawal of the hand and removal of the sensory stimulation. Internal balance is restored.

5.3.2. *Organizational Closure*

As was discussed in Section 3.1, the nervous system has a closed organization. As observers, we see a hot plate and a hand moving quickly away from it. It appears that the nervous system is an open system, receiving an input from the environment and producing an appropriate response. Yet, in view of Section 5.3.1, this account would be mistaken. The nervous system is in a process of continuous activity, the state of its components at one instant determining its state at the next. Thus states of relative neuronal activity are caused by and lead to further states of activity in an uninterrupted sequence. This seems clear for interneurons, which connect only to other neurons, but do not the sensory and effector surfaces constitute some kind of open interface to the world? Maturana argues that they do in an interactional sense, but not in an organizational sense.

The sensory surface is triggered by something in its environment, and its activity contributes to the activity of the whole. This may lead to motor activity compensating for the disturbance. The result is a further change to the sensory surface, not directly, but through the environment. The hand moves, the temperature falls. Relative activity leads to relative activity. This is equally true for sensory and effector surfaces that interact inside the organism. Excessive internal temperature leads to sweating and eventually to a restored temperature. In all cases nervous activity results from, and leads to, further nervous activity in a closed cycle.

Another way of saying this is that the nervous system is structure-dependent. Its possible and actual changes of state depend on its own structure at a point in time, not on some outside agency. At most, such an agency can only act as a trigger or source of perturbation. It cannot determine the reaction of the nervous system. This can easily be shown by recognizing that it is the structure itself that determines what can be a trigger for it. For instance, only systems with light-sensitive neurons can be affected by changes in light.

The points made in Sections 5.3.2 and 5.3.1 (i.e., the maintenance of internal correlations and organizational closure) apply equally to organisms that have no nervous system. That the amoeba is affected by certain chemicals, and that they lead to particular changes, is determined exclusively by the structure of the amoeba, not by the nature of the chemical. The next two points, however, are particular consequences of the nervous system.

5.3.3. Plasticity

A crucial feature of the nervous system is its *plasticity*—that is, the fact that its structure can change over time. This is because of the interneurons, which disconnect the sensory and motor surfaces, severing their one-to-one relations and vastly increasing the range of states open to an organism. It is this plasticity that allows changes in behavior, including that which we call *learning*. This plasticity affects not mainly the structure of connections between neurons and groups of neurons, but rather the pattern of response of individual neurons and their synapses. Such changes occur both because of the specific activity of interacting neurons and through the general results of chemical changes in the blood supply.

5.3.4. Interactions with Relations

Apart from introducing plasticity and thus change and development of behavior, the main feature of the nervous system is that it connects together cells that are physically separate within the organism. One vital result of this is that it allows the organism to act in response to the relations between events rather than the simple events themselves. An organism without a nervous system interacts only with isolated physicochemical occurrences. However, in organisms that have nervous systems connecting many different sensors, neurons develop that are triggered not by single events but by the relations that hold between events occurring simultaneously or, indeed, over time. Von Foerster (1984) gives an illustration of a network of neurons structured in such a way that it responds only to the presence of an edge—that is, a sharp discontinuity between light and dark.

This may well be the most important consequence of the nervous system. It enables organisms to interact with the general as well as the particular and leads to the possibility of abstract thought, description, and eventually language and the observer as will be shown in the next chapter.

6

Language and the Emergence of the Observer

6.1.   The Emergence of the Observer

The nervous system allows the relations that occur at the sensory surface to be embodied in a particular pattern of nervous activity. With the growth of the interneurons, this pattern no longer has a direct effect on the motor surface but constitutes a perturbation for the internal nervous system itself. The state of relative nervous activity becomes itself an object of interaction for the nervous system, leading to further activity. This is the basis for a further expansion of the cognitive domain, a domain of interaction with its own internal states as if they were independent entities. This is the beginning of what we term abstract thought.

The widened repertoire of behavior and the potential for change and development constituted a significant evolutionary advantage and stimulated an enormous expansion of the internal nervous system. Structurally, this development involved the nervous system projecting itself onto itself—the various sensory surfaces having corresponding areas within the cortex and these being functionally connected to one another and to various mediating structures. The human brain is vastly more responsive to its own internal structures than it is to its sensory/effector surfaces.

The next important emergence appears to be that of description and language. Maturana's ideas here are strikingly similar to those of G.H. Mead (1934) although apparently developed independently. The evolutionary developments outlined above lead to organisms with well-developed nervous systems capable of wide-ranging and adaptable behavior. Such organisms are structurally coupled to their environment and to other organisms within it. Complex sequences of mutually triggered behaviors are possible. Always, however, such behavior is, ultimately, structurally determined within each organism (Rosseel and van der Linden, 1990).

Within this context, Maturana distinguishes two types of interaction between organisms. In the first, the behavior of one leads directly to the behavior of the other, for example, fight/flight or courtship. The second is less direct. The behavior of the first organism "orients" a second organism, i.e., directs its attention to some other interaction that the two have in common. The orienting behavior stands for or represents something other than itself. What is important is that the behavior symbolizes something other than itself, and its success depends on the common cognitive domains of the organisms. This leads Maturana to describe the domain of such behaviors as *consensual,* and the interactions as communication.

Orienting behavior is thus symbolic; its significance lies not in itself, but in what it connotes or implies. In a very crude way, such a behavior is an action that is a description of the environment of an organism. It is the basis for the emergence of a new domain of interactions—the domain of descriptions—which in turn forms the basis of language. Initially these symbolic gestures are closely related, through metaphor and metonymy (Wilden, 1977), to the activity that they connote. However, the nervous system can interact with the corresponding states of neuronal activity as if they were independent entities and thus generate descriptions of descriptions in an endlessly recursive manner. In this way the symbols become further removed from their origin, and the domain of essentially arbitrary signifiers that we call language emerges.

As a result of this process and a concomitant development of the neocortex, organisms have arisen that can make complex and recursive descriptions of their descriptions, and thus they become observers. Moreover, within this linguistic domain a description of the self is possible, and thus descriptions of the self describing the self and so on. So is born the self-observer and self-consciousness.

To summarize Maturana's views so far: Autopoietic systems are structurally coupled to their media. Their behaviors are based on implicit presumptions or inferences about their environment and are thus cognitive. A nervous system does not alter this basic situation, but does permit the emergence of wider realms of interaction, culminating in the self-consciousness of humans. Initially, the nervous system severs the direct connection between sensory and motor surfaces, allowing a wider range of changeable behaviors and interactions with relations rather than isolated events. Increasing encephalization (i.e., development of the brain) under evolutionary pressure widens the range of possible behaviors to include abstract thought, orienting behavior, and the domain of descriptions. Finally, descriptions of descriptions and descriptions of self through language generate the observer and self-consciousness. At each stage emerges a domain of new and different interactions—interactions with relations, with internal nervous activity, with descriptions, with descriptions of descriptions, and finally with self-descriptions. All are made possible by the underlying biology, but none are reducible to it.

> The linguistic domain, the observer, and self-consciousness are each possible because they result as different domains of interactions of the nervous system with its own states in circumstances in which these states represent different modalities of interactions of the organism.
> Maturana and Varela (1980, p. 29)

6.2.   Consequences of the Theory

6.2.1.   Nervous System and Organizational Closure

As explained in Chapter 2, an autopoietic system is organizationally closed and structurally determined; its changes of state depend on its own structure at a time and are not determined (although they may be selected) by events in the environment. The same is true of the nervous system, even though it itself is not autopoietic. Every state of nervous activity leads to and is generated by other such states. This is true despite the fact that the sensory/effector surfaces appear to be open to the environment. The correctness of this counterintuitive view will be illustrated by a number of examples. Further discussion occurs in Section 11.3, which examines Varela's later work on embodied cognition.

First, let us examine studies of color vision in pigeons by Maturana et al. (1968). It might be expected that there would be a direct causal relation between the wavelength of light and the pattern of activity in the retina, and that this in turn would create the experienced color. In fact, it was not possible to directly correlate light wavelength and neuronal activity. The same nervous activity could be generated by different light situations, while the same wavelength of light could lead to different experiences of color [this is practically illustrated in Maturana and Varela (1987), pp. 16–20. See also Thompson et al., (1992)]. However, there was a direct correlation between retinal activity and the experience of the subject. In other words, a particular sensory activity always generates the same experience even though it may be triggered by different environmental situations.

Second, consider the sensory and effector surfaces of the nervous system between which lies an environment. Imagine a very simple nervous system with one sensor connected to one interneuron connected in turn to one effector. If the effector were itself connected directly to the sensor, then the closed circular operation would be apparent. It is not, but neither are the other neurons in this simple system connected directly to each other. They are connected across a small gap—the synapse—which therefore forms the environment between each pair of neurons. Moreover, each neuron can be seen as having its own effector and sensor surfaces. In principle, therefore, the relations between the sensory and effector surfaces of the nervous system are no different from those between any two neurons. What is different is that we, as observers, stand in one environment and not the other, and it is not apparent to us that functionally it is just as if we are standing within one of the synapses.

Third, let us examine the idea that the environment does not determine, but only triggers, neuronal activity. Another way of saying this is that the structure of the nervous system at a particular time determines both what can trigger it and what the outcome will be. At most, the environment can select between alternatives that the structure allows. This is really an obvious situation, of which we tend to lose sight. By analogy, consider the humming computer on my desk. Many interactions, e.g., tapping the monitor or drawing on the unit, have no effect. Even pressing keys depends on the program recognizing them, and pressing the same key may have quite different effects depending on the computer's current state. We say, "I'll just save this file," and do so with the appropriate keys as though these actions in themselves bring it about. In reality the success (or lack of it) depends entirely on our hard-earned structural coupling with the machine and its software in a wider domain, as learning a new system reminds us only too well.

As adults we are so immersed in and successfully coupled to our environments that we forget the enormous structural developments ("ontogenetic structural drift," in Maturana's words) that must have occurred in us, although observing the helplessness of young babies quickly brings this home. It is still easy, however, to imagine that the environment has caused us to become adapted to it, but this is as mistaken as to believe that the existence of tree tops caused the development of giraffes.

6.2.2. *Perception and Intelligence*

Maturana's approach brings out characteristically novel insights into these domains (Maturana and Guiloff, 1980; Maturana, 1983). In both cases he asks not, what is this phenomenon as an entity or characteristic, but what is this as a process generating the observed phenomena?

His analysis of perception has been introduced in Section 6.2.1. The process of perception does not consist of our grasping or representing an objective world external to us. Rather, it involves the operations of a closed neuronal network, which has developed a particular structure of sensory/effector correlations through a history of structural coupling. For the observer who sees the organism and its environment in apparent harmony, it seems that the organism must be responding to perceived changes in the environment. But the internal situation is rather like a robotic production. Each robot (neuron) is programmed to perform its own specific actions in orchestration with the others. While these actions are coordinated there appears to be purpose and communication, but as soon as they become unsynchronized the resulting ludicrous spectacle reveals how fragile is this illusion.

Similarly, intelligence is normally seen as an objective property of a person or animal, like weight or strength, which can be measured in an objective way by, for example, solving problems or puzzles. Maturana argues that we must ask how behavior that observers call intelligent is generated. His answer is that it must be the result of a history of structural coupling with the environment and/or other organisms, and that therefore any behavior that is successful within a domain of structural coupling is intelligent. Intelligence is neither a property of the organism or some part of the organism, nor is it directly observable. The word *intelligence* connotes the structure resulting from coupling in various domains, and it is manifest only in particular instances of coupled or consensual behavior.

There are a number of implications. First, all cultures, as consensual domains of biologically successful behavior, imply equivalent although not identical intelligence in their members. Secondly, intelligence in general cannot be measured and certainly cannot be compared across cultures. IQ tests reflect only interaction with a particular culture and can record only the extent of an organism's coupling to that particular domain, and thus to the observer (test creator) who specifies it. They cannot therefore measure the organism's potential for structural coupling in other domains, or in general. Third, specific intelligence is not heritable, for it is developed in the ontology of a particular organism's coupling. At most one can say that the general capacity for coupling in a particular domain (e.g., the linguistic) is genetically dependent.

6.3.   Language as a Consensual Domain

Just as it is mistaken to believe that the nervous system operates by manipulating the environment, it is equally mistaken to view language as denotative, that is, as objectively indicating and pointing to an external world. Linguistic behavior is connotative. The observed communication of meaning and the practical efficacy of language do not reside in the words and terms themselves but reflects similarities in the organisms' structures developed through their history of interactions.

As explained in Chapter 3, organisms that interact repeatedly with one another become structurally coupled. They develop behaviors that reciprocally trigger complementary behaviors, and their actions become coordinated so as to contribute to the continued autopoiesis of each. Moreover, the particular behaviors or conducts are divorced from what they connote; they are symbolic and thus essentially arbitrary and context-dependent. They only work to that extent that they reflect agreement in structure, and this is what Maturana means by a domain of consensual action. They rely on a consensuality (rather than explicit consensus) among those involved (Harnden, 1990):

> When two or more organisms interact recursively as structurally plastic systems, each becoming a medium for the realization of the autopoiesis of the other, the result is mutual ontogenetic structural coupling. . . . The various conducts or behaviors involved are both arbitrary and contextual. . . . I shall call the domain of interlocked conducts that results from ontogenetic structural coupling between structurally plastic organisms a *consensual domain*. . . . A consensual domain is closed with respect to the interlocking conducts that constitute it, but is open with respect to the organisms or systems that realize it.
> Maturana (1978, p. 47)

The consensual domain is thus a domain of arbitrary and contextual interlocked behaviors. Much animal behavior involves coordinating actions of this type, e.g., courtship, nest-building. Some may be instinctive, e.g., the dance of bees, but most is learned through the structural drift of the organism through its life. This learned consensual behavior Maturana terms *linguistic*, although it is not yet language. It is distinguished by its symbolic nature—i.e., that the action stands for something other than itself. For an observer, such coordinating conducts can be seen as a description of some feature of the organism's environment.

Linguistic acts by themselves do not constitute language. For Maturana, the process of using language, or *languaging*, can occur only when the linguistic behaviors themselves become an object of coordination. This in turn can happen only when the nervous system has developed in such a way that it can interact with its own symbolic descriptions. Thus linguistic behavior is the consensual coordination of action. Languaging is a recursion of this, i.e., the consensual coordination of consensual coordinations of action.

Once this level of abstraction has been reached—i.e., the description of a description—the entire space of language is opened up, as is the observer and the self-conscious self-observer. In his early work Maturana talked of descriptions and descriptions of descriptions, but now he refers to consensual coordination of action. This emphasizes his view that language is not essentially a descriptive domain but always an activity, embedded in the ongoing flow of actions.

Having uncovered the genesis of human language, let us move to the level of its day-to-day use. Maturana (1988b) has developed an elegant description of languaging around the concept of a conversation—that is, an ongoing coordination of actions in language among a group of structurally coupled observers. For the individual, such a conversation is actually a meshing or braiding of language and mood (or emotion). The linkage between these distinct domains occurs because they are both embodied in the body of the observer. Although often ignored in discussions of language and meaning, in real conversations our mood or *emotioning* is an ever-present background to our use of language. It conditions our stance or attitude (are we happy or sad, caring or self-concerned, deferential or confident, angry or upset?) and thereby the course of our conversation. In turn, what we say and what is said to us may trigger in us changes of mood. For Maturana a conversation is an inextricable linking of language, emotion, and body, and the nervous system is the medium in which all intersect.

As Winograd and Flores (1987a) have recognized, this view is strikingly similar to that of the phenomenology of Heidegger (1962). He too argues that in relating to the world, in existing in the world, our basic attitude is always a practical one of doing, of acting, of having some aim in mind. Our consciousness (although we may not generally be conscious of this) is characterized by our state of mind or mood and by our understanding of our situation, which may be articulated in language. Generally, we are immersed in our daily tasks and do not notice most of the world as such. In using language within a conversation, we bring out particular objects and highlight particular properties in the light of our concern at the time. This will be considered in detail in Sections 7.4 and 7.5.

It is important to note that the driving force behind these developments is the evolutionary advantage they provide by permitting cooperative and coordinated activity. Thus language itself is ultimately rooted in cooperative practical activity and its effects, rather than the abstract exchange of meaning and ideas. It also emphasizes that language is itself an activity and of course is not restricted to verbal actions alone. It is interesting to compare this with Habermas' analysis of language.

For Habermas (1979), too, language is a practical activity that arises out of the need for the social coordination of action. This idea has important consequences for the underlying nature of language, namely that for utterances to be practically successful they must make, at least implicitly, certain claims as to their validity. Over and above being comprehensible, they must be true in their description of the external world, right according to the norms of the social world, and truthful in their expression of the subjective world of the speaker. Habermas, although accepting the intersubjective nature of language, remains wedded to the denotative view of language and communication.

6.4. Typologies of Systems

There are many different ways of classifying types of systems, for example, Boulding's (1956) hierarchy of complexity, Jordan's (1968) systems taxonomy, and Checkland's (1971) systems map of the universe. In this section I will show how Maturana's and Varela's work can provide a fruitful reconceptualization of Boulding's hierarchy, and I will also consider different types of "self-referential" systems.

6.4.1. A Reconceptualization of Boulding's Hierarchy

Boulding offers an intuitive classification of different types of systems ranked in increasing order of complexity (Table 6.1). This is done by

TABLE 6.1
Boulding's Hierarchy of Complexity

Level	Characteristic	Example	Discipline
1. Structures and frameworks	Static	Bridge, mountain	Descriptive
2. Clockworks	Predetermined motion	Clocks, machines, solar system	Physics, natural science
3. Control mechanisms	Closed-loop control	Thermostat, homeostasis	Cybernetics
4. Open systems	Structurally self-maintaining	Flames, cells	Theory of metabolism
5. Lower organisms	Organized whole, functional parts	Plants	Botany
6. Animals	A brain, ability to learn	Birds, mammals	Zoology
7. Man	Self-consciousness, knowledge, language	Human beings	Biology, psychology
8. Sociocultural systems	Roles, communication, values	Families, boy scouts, clubs	History, sociology, anthropology
9. Transcendental systems	Inescapable unknowables	God?	Theology

specifying a number of different levels, characteristics, and examples of each, as well as relevant academic disciplines. At first sight, the levels and examples seem very plausible. However, as Checkland (1981, p. 106) has noted, there is no explicit definition of the scale of "systems complexity" underlying the hierarchy. What is it that actually distinguishes one level from another? Without this, we cannot really judge the correctness of the classification. I propose an answer to this problem, based on Maturana's and Varela's work. I suggest that the underlying dimension is the way in which the system's organization is characterized by *different types of relationship*. Each new level brings in a different relation, or relation of a relation, as well as involving those at previous levels.

Table 6.2 shows the hierarchy reconceptualized in these terms. Most levels are the same, but certain inconsistencies in the original are cleared up. At Level 1 we have structures and frameworks characterized

TABLE 6.2
A Hierarchy of Complexity (Developed from Boulding)

Level	Characteristic	Type of relations	Example	Domain
1. Structures and frameworks	Static	Topology (constitution)	Bridge, mountain	Mechanistic systems
2. Simple mechanistic systems	Dynamic, open	Order	Solar system, clock, flame	Dynamic systems
3. Self-regulating systems	Negative feed-back	Specification	Thermostat, body temperature system	
4. Living systems	Self-production	Autopoiesis	Cell	
5. Multicellular system	Second-order autopoiesis	Structural coupling between cells	Plants, fungi	
6. Organisms with nervous systems	Closed organization	Interaction with relations	Animals (except sponge)	Linguistic domain
7. Observering systems	Language, self-consciousness	Relations of relations	Humans	Consensual domain
8. Social systems	Third-order autopoiesis	Structural coupling between organisms	Families, ant colonies	Social domain

by relations of *constitution* (see Section 2.3.2), that is, spatial or topographic relations. Level 2 I call "simple mechanistic systems." These are dynamic and involve relations of *order*. Unlike Boulding, I include simple open systems, such as candles, here. Boulding put these in the same level as cells which, in the light of autopoiesis, is clearly wrong. Level 3 consists of self-regulating systems such as a thermostat or the body's temperature-regulation system. These systems maintain certain quantities as constants through negative feedback. Because of this, they are characterized by relations of *specification*. That is, they maintain variables at *specific* levels.

At Level 4 we reach living systems whose organization is characterized by relations of *autopoiesis*. That is, a relationship among the relations of typology, order, and specification. This replaces Boulding's Level 4 of open systems, which incorrectly included both flames and cells. Level 5 consists of multicellular systems, which involve relations of *structural coupling between cells*—what Maturana calls "second-order autopoietic systems," such as plants and fungi. This is a more precise definition than Boulding's "lower organisms." Level 6 consists of organisms that have nervous systems which allow them to interact with pure relations—that is, relations between events rather than isolated events alone (Section 5.3.4). This category includes all animals except the sponge. Again, Boulding was inconsistent in using plants as an example, but animals as a level.

Level 7 brings in systems which can observe, that is, make descriptions and self-descriptions. This involves specifying *relations between relations* and requires language. Note that Level 6 animals can have *linguistic interactions* such as a duet of birdsong, but Maturana argues that this is not full languaging (Section 6.3). The main example is human beings. Finally, we have Level 8, social systems. These are characterized by relations of *structural coupling among organisms*—that is, organisms coordinating their behavior with respect to others. Examples are social insects such as ants, animals such as gorillas, and humans, with our social systems such as families (Section 8.4). Note that Maturana does not claim that social systems are autopoietic, but that they are the *medium* for the operation of autopoietic systems.

6.4.2. Classifying Self-Referential Systems

Autopoiesis focuses our attention on self-producing, self-referring systems, but as Teubner (1993, p. 16) has noted, there is considerable conceptual confusion around such terms in the general systems literature, and even, on occasions, within Maturana's and Varela's work. What is the difference between, for example, self-reference, self-production, self-organization, self-regulation, self-observation, circular causality, tautologies and contradictions, and autocatalysis? While not claiming a complete analysis of such systems, I offer the categorization in Table 6.3 as a start.

The first type I identify as *self-influencing* systems. These are systems including what are often called causal loops or circular causality— that is, patterns of causation or influence that become circular (for instance, the larger a population, the greater the number of births, and thus the larger the population). This creates a positive loop leading to exponential increase or decrease and, more commonly, there are negative loops which lead to stability. The second type are *self-regulating* systems, which are organized so as to keep some essential variable(s) within particular limits. They rely on negative feedback and specified limits. The next type I call *self-sustaining* systems. These are systems that are organizationally closed but not self-producing. Their operations close in upon themselves so that they are necessary and sufficient for their continuance. A good example is the gas-heater pilot light discussed in Section 3.3. Here, the pilot light heats the thermocouple, which controls the gas, which allows the pilot light to function in the first place. Once started, it sustains itself.

After this we have autopoietic systems, which are *self-producing* of both their components and a boundary. Next come *self-referential* or *self-referring* systems. To avoid confusion, I restrict these specifically to sym-

TABLE 6.3

A Classification of Closed Systems

Type	Characteristic	Example
Self-influencing	Circular causality, causal loops	Size and birth rate of population
Self-regulating	Maintenance of a particular variable	Thermostat, body temperature
Self-sustaining	Organizational closure but not self-production	Gas pilot light, autocatalysis
Self-producing	Autopoiesis	Cell
Self-referential (self-referring)	Symbolic reference to self (pictorial or linguistic)	"This is a sentence," Escher's "Drawing Hands," Magritte's "The Treason of Images"
Self-conscious	Able to interact with descriptions of self	A person saying, "I acted selfishly today"

bolic systems that can make reference to themselves. These are generally linguistic or pictorial and involve all the many paradoxes and
tautologies (Hughes and Brecht, 1978). Pictorial examples are Escher's
drawing ("Drawing Hands") of two hands drawing each other and Ma-
gritte's pictures, one ("The Treason of Images") of a (smoking) pipe with
the words (in French) "this is not a pipe," and another ("The Human
Condition") in which the picture contains a picture of part of the scene
in the larger picture. Finally, there are *self-conscious* systems, which can,
through language, create descriptions of themselves and then interact
with these descriptions, thus recursively generating their conscious
selves.

One noticeable omission is a category of *self-organizing* systems.
This seems to me redundant within Maturana's terminology. Each systems embodies a particular *organization*—it is what specifies the system's
identity—and this organization remains the same so long as the system
maintains its identity. What is implied by self-organization is actually
structural change, and here I would use my other categories, depending
on the particular type of change involved.

7

Philosophical Implications

7.1.   Introduction

It is especially important that we consider the philosophical implications of Maturana's work. All theoretical work in the natural and social sciences has philosophical implications in that, explicitly or implicitly, it embodies particular epistemological and ontological presuppositions. However, most work does not produce results which, in themselves, have philosophical implications. This is not the case with autopoiesis and the allied theories of cognition since they constitute, *inter alia,* a theory of the observer. That is, they claim to define the possibilities and limitations, biologically, of cognition and therefore of observation, description, and knowledge. This, of course, makes them highly self-referential, for the results must apply equally to the theory that generated them. This, in itself, does not cause a contradiction, but it does place a premium on consistency. Maturana himself is very well aware of all this, and has continually made explicit what he sees as the radical philosophical implications of autopoiesis.

These cognitive theories lead to an antirealist position, according to which we can have no claim to *objectivity,* in the sense of subject-independent truths—we have no access to an objective reality. Our beliefs and theories are purely human constructs which *constitute* rather than *reflect* reality. This position has been termed *constructivism.* This chapter seeks not only to explain Maturana's developing position but also (unlike previous chapters) to critique it, since it seems most appropriate to link the two activities together. Section 7.2 provides a brief overview of the main positions within the philosophy of science—realism, empiricism, and idealism. Section 7.3 explains the main ideas as recorded in the early works on autopoiesis and the nervous system, and Section 7.4 presents his ideas as developed in later work. Section 7.5 concludes that Maturana's radical position cannot be maintained and that his work is best seen as an example of the recent development of critical realism as seen in the works of Bhaskar (1978, 1989).

7.2.   An Outline of the Central Philosophical Worldviews— Empiricism, Realism, and Idealism

Maturana contrasts his own views with those which he claims "are generally held by scientists, both explicitly and implicitly" (1974, p. 457), namely that

1. We exist in an objective world about which we can make valid statements that are independent of us as observers.
2. We can gain valid knowledge of this world through our sense organs, and the result is representations of reality that we use to guide our behavior.

This view might best be termed the naive realist or common-sense view and may characterize the views of unphilosophical scientists. Within the philosophy of science, however, there are a range of sophisticated epistemological positions, none of which involves such a simpleminded objectivism. This section presents a simplified sketch of some of the main epistemological traditions in order to facilitate a discussion of Maturana's work. It draws on Outhwaite (1987), Hamlyn (1987), Harre (1986), Keat and Urry (1982), and Bhaskar (1989).

If we distinguish among three possible domains—the world independent of ourselves, our perceptions of the world, and our ideas and theories—then we can distinguish three broad paradigms with a long history in philosophy: realism, empiricism, and idealism. These differ centrally in placing primary importance on the world, on our perceptions, and on our ideas, respectively. While empiricism is essentially epistemological, realism and idealism also have ontological implications.

Empiricism, beginning with Locke and Hume and culminating in positivism, has been the dominant view in the Anglo-Saxon world. It was a reaction against the grand religious and rationalist systems and holds that valid knowledge must be based on our observation and experience of the world as we find it, rather than, on the one hand, abstract rational or introspective ideas (rationalism and idealism) or, on the other hand, unobservable causes or theoretical entities (realism). Classical empiricism developed from Hume's view of causation as purely a constant conjunction of events.* In this view, scientific laws and theories simply connect up the regularities that occur within our observations and perceptions rather than explaining why they occur. Indeed, phenomenalists hold the view that nothing other than the phenomena perceived by our senses exists. This tradition culminated in *positivism* and *behaviorism*—holding that scientific, and indeed rational, knowledge must be based solely on empirical facts. These would be explained as examples of universal laws developed as hypotheses and confirmed by prediction and experimentation. The individual subject or observer is merely a passive reflector or channel for the phenomena of reality.

Idealism in its many forms has an even longer history. It emphasizes the importance of ideas and thought either in shaping our individual experience of the world (e.g., Kant) or as existing in themselves (Plato, Hegel). Recent developments in this tradition have effectively destroyed the empiricist belief in the fundamental role of observation and focused attention on the role of the subject or community of subjects in developing knowledge.

This has occurred at a number of levels. First, it has been shown that our perception of the world is very much an active construction rather than a passive reception of sense data and thus reflects us as much as the world (Hanson, 1969). Second, it has been argued that scientific observation is theory-dependent; that is, our theories determine what experiments to perform, what instruments to construct, how to interpret the results, and generally how to conceptualize the world (Kuhn, 1970; Feyerabend, 1975; Popper, 1972). No experience or observation can be free of our presuppositions. Third, particular theories or statements cannot be evaluated in isolation but are part of complex paradigms (Kuhn, 1970), which are essentially incommensurable. Fourth, science is essentially a social activity, and scientific development takes place through the judgment of the scientific community (Woolgar, 1988). These various, quite radical, critiques of the objectivity of empiricism have generated no single dominant view as to the nature of scientific theories. I shall describe certain positions under the terms *conventionalist* and *constructivist*.

Conventionalism argues that scientific theories cannot be wholly determined by the nature of the world and our observations. The choice of theory is to some extent subjective and conventional. Two particular

---

*That is, that in saying A caused B we are not making a positive assertion about an operative relation, merely noting that the prior occurrence of A has always been associated with the occurrence of B.

connected examples are *pragmatism* and *instrumentalism*. Pragmatism, as developed by Peirce (Buchler, 1955), James (1948), and Dewey (1931), pointed out that science is essentially a practical activity, both in itself and in being useful in solving people's problems. The meaning of terms comes from their use or the effects that they have, and statements are judged true not because they correspond to some external reality but because they solve the problem or cure the doubt at the time. Instrumentalism (Duhem, 1954; Rapoport, 1969) also denies that theories correspond to the world or have representational meaning, arguing that they should be seen as purely predictive devices or instruments, improving our manipulative power over the world.

Constructivism (Von Glasersfeld, 1984; Boyd, 1984) refers to the more radical idea that our theories and, indeed, our experiences of the world are essentially constructed by us; we construct the world that we experience, either as individuals or as communities, and as our theories change so does the world we experience. Such a view can be inferred from Kuhn and is common in social theory (Berger and Luckmann, 1967) and phenomenology (Heidegger, 1962).

Realism, in naive form, echoes our common-sense experience. There is an objective, independent world, which we experience directly; our theories describe actually existing entities; they are true or false by virtue of corresponding to the real world. In this form realism has rightly been subject to the idealist criticisms outlined above. However, in recent years a more sophisticated version has developed (Harre, 1970; Bhaskar, 1978, 1979, 1989; Keat and Urry, 1982; Leplin, 1984). This new realism, called by Bhaskar *transcendental* or *critical realism*, accepts the epistemological criticisms that observations are theory-dependent and that we cannot have pure access to an independent world. It asserts strongly, however, that such a world does exist and that it is populated by objects and entities, some of which may be in principle unobservable, which have causal powers or tendencies. Against empiricism, it argues that it is the interaction, in complex ways, of these generative mechanisms that leads to our observations. Science can no longer be seen as creating true theories, but as proposing and identifying potential causal objects, the descriptions of which are at least approximately true.

The methodology of critical realism, described as "retroductive," is (Bhaskar, 1979, pp. 144–148):

1. Identify an effect or phenomenon to be explained.
2. Postulate a hypothetical mechanism or structure that, if it existed, would generate the phenomenon.

3. Attempt to demonstrate the existence of the mechanism by experimental activity leading to direct or indirect observation and by eliminating alternative explanations.

7.3. The Philosophical Conclusions of Maturana's and Varela's Early Works

This section will introduce the philosophical positions contained in the early work of Maturana and Varela. (Section 7.4 will consider in detail the later work developed by Maturana.)

7.3.1. Introduction

Unlike many scientists, Maturana and Varela have always been very aware of the philosophical implications of their work, and have devoted much time to making them explicit in their own terms. However, as in their biological work, Maturana and Varela develop their own language and there is little attempt to relate it to already existing concepts and positions. A main aim of this book is to try to translate their ideas into more familiar terms, although I recognize that Maturana particularly might object that any such translation will inevitably be distorting. Put crudely, Maturana and Varela argue that our descriptions and theories are human constructs, which reflect our individual and communal subjectivity rather than an independent objective world.

Maturana began his work by pursuing two questions: what is the nature of living organisms, and what is the nature of perception? His breakthrough was to see that these were in fact intimately connected. Initially explored in *The Biology of Cognition* (1970b), this led to a characterization of the observer as the system in which description takes place. As humans, we are autopoietic systems, each with a plastic nervous system that is organizationally closed. This is structurally coupled to the rest of our body and through this to the environment. Both the body and the nervous system are structure-determined systems; the changes they undergo depend on their own prior structure and can only be triggered, not determined, by interactions with other systems. The capacity for language has opened up new domains of activity in which we may make descriptions, descriptions of descriptions, and descriptions of ourselves, leading to the emergence of the observer. However, such descriptions always remain subject-dependent and based on consensus.

Moreover, science and scientific theories are inevitably expressed in some form of language (including that of mathematics) and therefore lie within the domain of descriptions—the consensual domain. Such descriptions are not determined by the nature of what is described, but by the describer. They cannot, therefore, reflect an objective reality; they must remain subjective constructions. This has clear implications for ontology, epistemology, and methodology.

In this section Maturana's early ideas will be expounded, largely in his own words.

7.3.2. *Ontology, Epistemology, and Methodology*

7.3.2.1. Ontology

Ontology deals with the nature of existence. That is, what kinds of objects and entities do we take to exist in the world and what are their modes of existence? In other words, what might be the object of our descriptions? Here we find Maturana's most radical assertions, which can be put forward in the following argument:

A. All our perceptions and experiences occur through and are mediated by our bodies and nervous systems. It is therefore impossible for us to generate a description that is a pure description of reality, independent of ourselves.

> No description of an absolute reality is possible. Such a description would require an interaction with the absolute to be described, but the representation which would arise from such an interaction would necessarily be determined by the autopoietic organization of the observer, not by the deforming agent.
> Maturana and Varela (1980, p. 121)

This latter point is because of B:

B. The structure-determined nature of the observer means that interactions are selective rather than instructive, so that the resulting experience always reflects the observer.

> . . . the environment, . . . as an independent domain of perturbations, can only participate in the selection of a particular sequence of states along the ontogeny of the organism, from the domain of states that at each instant the structure of the organism makes possible.
> Maturana (1975b, p. 11)

C. We are therefore always, unavoidably, existing within a domain of subject-dependent descriptions—that is the only reality we can experience—the reality of the distinctions and descriptions that we make. We must accept this and alter our view of our world accordingly.

> . . . reality as a universe of independent entities about which we can talk
> is, necessarily, a fiction of the purely descriptive domain. . . . we should
> in fact apply the notion of reality to this very domain of descriptions in
> which we, the describing system, interact with our descriptions as if with
> independent entities.
> Maturana and Varela (1980, p. 52)

D. Thus the object of our knowledge does not exist except as it is
distinguished by the observer. It is not just that we cannot access an
existing external reality, but that *our realities* are brought into existence
through our activities as observers.

> The question, "What is the object of knowledge?" becomes meaningless.
> There is no object of knowledge. . . . we, as thinking systems, live in a
> domain of descriptions. . . . Furthermore, this re-emergence of reality as
> a domain of descriptions does not contradict determinism and predicta-
> bility. . . .
> Maturana and Varela (1980, p. 53)

E. This is true not just of the everyday world of ordinary people,
but also of the world of science itself, which is equally confined to exploring the results of its own distinctions.

> A universe comes into being when a space is severed into two. A unity is
> defined. The description, invention, and manipulation of unities is at the
> base of all scientific inquiry.
> Maturana and Varela (1980, p. 73)

7.3.2.2. Epistemology

Epistemology concerns the nature of our knowledge about the world.
How is knowledge (*episteme*) different from mere opinion (*doxa*)? What
are the criteria of validity for knowledge? Can knowledge be objective,
and what does that mean? What are the limits of knowledge? Given
Maturana's ontology, it appears that we can have no access to an independent world with which to compare our theories and descriptions.
How, then, are we to judge their truth or validity? And how, indeed,
are we to judge the validity of Maturana's theory itself? Maturana gives
a number of rather different answers to the question:
    A. The validation of knowledge is the maintenance of successful
autopoiesis. False knowledge leads to the destruction of the autopoietic
process. This is because, for Maturana, all activity is cognitive activity
and as such makes implicit presumptions about the world.

> Successful interactions directly or indirectly subservient to the mainte-
> nance of his living [autopoietic—JM] organization constitute his only final

source of reference for valid behavior within the domain of descriptions, and hence, for truth.
Maturana and Varela (1980, p. 57)

To know is to be able to operate adequately in an individual or cooperative situation.
Maturana and Varela (1980, p. 57)

B. Scientific statements are validated by the scientific method, which is used to produce them, not by correspondence to an external world. Proper scientific methodology includes specifying the observer, or rather the observer as a constitutive member of a community of observers.

We as scientists make scientific statements. These statements are validated by the procedure we use to generate them: the scientific method. . . . Science is necessarily a domain of socially accepted operational statements validated by a procedure that specifies the observer who generates them as the standard observer.
Maturana (1978a, p. 28)

The validity of scientific knowledge rests on its methodology, which specifies the cultural unity of the observers, not in its being a reflection of an objective reality, which it is not.
Maturana (1974, p. 464)

C. There is one thing common to all domains, and that is logic itself. This must apply to both the description and that which is described.

In every explanation . . . the reformulation of the phenomenon to be explained resorts to the same notions (identity, exclusion, succession, etc.). There is, then, a universal logic valid for all phenomenological domains . . . and the validity of our arguments, as the validity of any rational argument or concrete phenomenological realization rests on its validity. . . . To the extent that we have been successful (free from logical and experiential contradictions), we can conclude that [the physical space is one in which] the logic that we have applied in our description is intrinsically valid.
Maturana and Varela (1980, p. 121)

7.3.2.3. Methodology

As we can see from B above, methodology is important for Maturana, as it is a major determinant of the validity of a scientific theory. He suggests that there are four stages (Maturana, 1978a, p. 28):

a) . . . observation of a phenomenon that, henceforth, is taken as a problem to be explained;

b) proposition of an explanatory hypothesis in the form of a deterministic system that can generate a phenomenon isomorphic with the one observed;

c) proposition of a computed state or process in the system specified by the hypothesis as a predicted phenomenon to be observed; and

d) observation of the predicted phenomenon.

Stated thus, it sounds unoriginal and objectivist, and does not really reflect the thrust of his subjective, subject-dependent position. We shall see later how this basic formulation is developed to become more consistent with the rest of the theory.

7.3.3.   Preliminary Analysis of Maturana's Philosophy

It seems clear that the whole thrust of Maturana's work places it centrally within the idealist tradition, and more specifically what I called *constructivist*. In fact, I shall argue in Section 7.5 that there are remarkable similarities to the phenomenology of Husserl and Heidegger despite the fact that these authors have never been mentioned by Maturana [although they have by Varela (Varela *et al.*, 1991)]. However, it is also possible to pick out resonances with many of the other positions described—in particular, pragmatism and (implicitly) realism.

Maturana's theory provides a coherent explanation (or generative mechanism, in his terms) that blends a number of philosophical ideas. But for the main distinctions made in Section 7.2, it is clear that he is neither an empiricist, in that he maintains that experiences are not independent of the observer, nor a realist, denying access to or the existence of independent reality. He argues that what primarily exists for us human beings are the phenomena of our experience rather than an independent reality (phenomenalism). However, these phenomena are not necessarily the same for all, but are subject-dependent, generated by the operation of a structure-determined but plastic nervous system within a consensual domain. As such, we construct the world we experience (constructivism). These constructions are not purely individual, but reflect the intersubjective nature of language and action. Different domains of experience give rise to different domains of reality. As observers, we describe and explain our experiences as part of our practical daily existence, and our explanations are judged valid if they satisfy listeners according to the criteria appropriate to their domain, rather than by virtue of being true or false (pragmatism).

After illustrating this in a more detailed description of Maturana's recent developments, I shall argue, against Maturana, that his work is compatible with a critical realist approach and that it should be thus interpreted.

7.4. *Maturana's Developed Position*

The central tenet of Maturana's ideas—that the world we experience is a subject-dependent creation—was apparent in his earlier works as out-lined in Section 7.3, but one of the most developed and comprehensive expositions is a paper entitled "Reality: The Search for Objectivity or the Quest for a Compelling Argument" (Maturana, 1988b). An expanded discussion of the nature of science and scientific explanation is given in Maturana (1990).

In Maturana (1988b) he begins by adopting a very pragmatic view of explanation and its role in our lives—what he refers to as the *praxis of living*. As humans, we are linguistic animals; all our thoughts and experiences are mediated through language. And much of our linguistic activity consists of explanations, to others or to ourselves, of our expe-riences. It is important to recognize, however, that such explanations are always *post hoc*—after the event. We are always already experiencing and acting in the world before we ever question or explain an experi-ence. This means that explanations are always secondary to the actual praxis of living, yet they occur only within it and they feed back to affect ongoing behavior. Moreover, explanations are always made by observ-ers to other observers, who must accept or reject them; the circle can never be broken.

What then is the explanation for explanations? What motivates an explanation and then makes it acceptable to another person in daily life? Not, Maturana argues, its intrinsic truth or correspondence to an inde-pendent reality but simply whether or not it satisfies the listener accord-ing to whatever criteria the listener finds appropriate within his or her own praxis—whether it makes the listener content. A question is asked. An explanation is given. The explanation is always the reformulation of an experience in our praxis of living and must be accepted by the listener (who may also be the speaker) through the listener's own praxis of liv-ing. We never hear explanations purely in their own terms; we always react to them in accordance with our own criteria of acceptance, our own *pre-judices*.

7.4.1. *Paths of Explanation*

Maturana then distinguishes between two fundamental ways of vali-dating or accepting explanations (or, as he calls them, paths of expla-nation): "objectivity-without-parenthesis" *(transcendental objectivity)* and "objectivity-in-parenthesis" *(constituted objectivity)*. These two paths (or *explanatory domains*, as he also calls them) are primarily epistemological, but they also have ontological implications; indeed, he uses the terms *transcendental ontology* and *constitutive ontology*. Each domain is characterized by criteria for validating explanations, assumptions about the nature of the entities involved, and actions (and therefore cognition) that are seen as legitimate:

The path of transcendental objectivity corresponds essentially to the realist view of reality. According to Maturana, the observer following this path accepts that much of reality takes place independently of the observer's activities; that things exist independently of the observer knowing them; and that knowledge can be gained through perception and reason. In this domain, explanations are held to gain their validity by their reference to an independent reality, which is the criterion of acceptability. They are thus "transcendental" in that they precede or are prior to the observer. The active participation of the observer in the constitution of reality is overlooked, and observers require that there be and accept that there is a single domain of reality. To the extent that observers actually generate their own individual domains of reality, these must of necessity be exclusive, and those of the others negated. In this path, therefore, observers claim that their own views have privileged access to objective reality, using this as a justification for their choices and avoiding personal responsibility.

The path of constituted objectivity places objectivity or independent reality in parentheses. It recognizes that we are living systems and that our cognitive abilities have a biological base. Therefore, we cannot assume that our perceptions correctly represent some independent environment. Neither, in neurophysiological terms, can we distinguish between reality and illusion. As we cannot access an independent reality, we should suspend our naive belief in it. Instead, we should acknowledge that existence for us is constituted by us through our linguistic distinctions. This is what is meant by "constitutive ontology"—we can interact only through linguistic entities, and they come into being as they are constituted by us.

Exploring this view, Maturana characterizes it as follows:

> In the path of objectivity-in-parenthesis, existence is constituted by what the observer does, and the observer brings forth the objects that he or she distinguishes with his or her operations of distinction as distinctions of distinctions [*sic*] in language. Moreover, the objects that the observer brings forth in his or her operations of distinction arise endowed with the properties that realize the operational coherences of the domain of praxis of living in which they are constituted. [This path entails] the recognition that it is the criterion of acceptability that the observer applies . . . that

determines the reformulations of the praxis of living that constitute explanations in it. . . . Each configuration of operations of distinctions that the
observer performs specifies a domain of reality.
Maturana (1988b, p. 30)

We as observers always operate in and through a domain of language, that is, a domain of consensual verbal and nonverbal behavior
that intersects with the other domains of our experience. In acting
within our praxis of living, we create or construct distinctions and categories—and thus *bring forth* the objects of our language. We do so for
many different and nonintersecting domains, and each domain has its
own particular objects and criteria of validity. Examples of such domains
might be physical existence, poetry, music, games and sports, and science, each of which is a complex edifice of conventions and distinctions
erected on its own self-referring foundations.

Each domain is characterized by particular operational coherences,
which generate the characteristics of the objects we construct in that
domain, and each is a domain of explanations created by reformulating
our experiences in that domain according to its operational coherences.
These domains are different but all equally legitimate, so there may be
conflicting yet valid explanations across different domains. The observer
lives in a "multiverse [of] many different, equally legitimate, but not
equally desirable, explanatory realities" (Maturana, 1988b, p. 31). If we
accept this path and recognize the validity of other people's domains,
then we must be doing so under conditions of mutual acceptance rather
than conditions of mutual negation. Maturana characterizes the emotion
underlying mutual acceptance as "love" (in a nonphysical or spiritual
sense) and claims that it is the basis of social phenomena (Maturana,
1988b, p. 64). This will be discussed further in Chapter 8.

7.4.2. *Operational Coherences*

Before looking in more detail at these explanatory domains, it is necessary to discuss the term "operational coherences." This is used repeatedly by Maturana in recent works without any very clear explanation of
its precise implications. Indeed, it seems to play a crucial role both in
his explanation of daily life (as above) and in his characterization of the
nature of science (Section 7.4.4).

What is Maturana trying to get at with this enigmatic phrase? It is
important to reiterate some of the fundamental premises of his whole
position, starting with the intimate bond between knowledge and action. (Indeed, they are really just different ways of looking at the same
thing.) Knowledge is effective action; cognition is action that continues

and maintains autopoiesis. And we must remember that action is not just observable behavior, but includes any activity of the body or nervous system, such as thinking or reflection. Second, as biological systems we are structurally determined; all our actions and experiences must exist in a closed world of self-reference, continually perturbed but not determined by outside events.

Third, we exist in language and, through language, both experience and explain our experiences by bringing forth the particular distinctions that we do. We can communicate these to others because of the history of structural coupling between us which leads us, within a particular domain, to create mutually agreed upon distinctions. Finally, our physiology operates in such a way as to maintain internal relationships, that is, it "seeks" recurrence [i.e., consistence or coherence] (Maturana and Varela, 1987, p. 231). This occurs at all levels of interactions—physical autopoiesis, the closed nervous system, the self-descriptive domain of the observer, and the consensual domain of language.

Thus, in producing an explanation we specify a particular entity, mechanism, or process and implicitly or explicitly define the way it operates as a structure-determined system. We do this on the basis of our own subject-dependent experiences in language, not as an objective description of something independent of us. We define the *operational characteristics* or *coherences* of the entities as we expect that we would experience them. Other people will accept or reject these characterizations to the extent that they share our domain of experiences. The point about it is that these distinctions and descriptions form a closed set, building one upon the other but undetermined by an external world in just the same way as a physical autopoietic system is closed organizationally. If we ask why they appear to "work" so well, the answer is also the same—structural coupling based on effective action. If they did not "work," if they were not effective, they would not continue self-producing.

Let us look at an example. A colleague walks into my office and asks, "what's keeping you busy?" I point to a pile of papers and reply," exams to grade." She smiles sympathetically and walks out. A question has been asked and an explanation given and accepted. Within the domain of education, which we share, I have brought forth a particular entity, "exams to grade," as a mechanism capable of generating the experienced phenomena, my apparent business. These three words carry with them a whole raft of connotations concerning the exam system, students, writing exam questions, marking schemes, deadlines, boredom with grading, and so on. These constitute the operational coherences entailed by giving and grading exams.

It might be thought that these are objective entities and character-istics available to anyone, but this is not so. They can only be construc-tions or reformulations of *my* experience, although they may be similarly experienced by others who share this domain. They are different for those who do not share this domain. For students, "exams to grade" entail quite different operational coherences—revision, anxiety, past pa-pers, results, and so on. For someone from a completely different cul-ture, this explanation may make no sense at all. The explanation is successful because it corresponds to or could generate the observed ex-perience shared by my colleague and me, not because it corresponds to objective reality.

7.4.3.   Domains of Reality

In the path of objectivity-in-parenthesis, it must be accepted that there are many different, equally valid, domains of reality. Each domain is constituted in three interlocking dimensions—the criteria for accepting explanations, different operational coherences structuring such expla-nations, and the actions seen as legitimate. Together, these define a cog-nitive domain—a domain of possible viable existence. In our daily lives we experience many such domains, constantly switching from one to another. Each is equally rational and equally consistent, and to the ex-tent that we can choose between them the choice depends only on our preference—our emotional preference. Maturana emphasizes that emo-tion or mood is an ever-present condition of our experience in the world (see Section 6.3).

For example, in the domain of religion, there are many objects, doc-trines, practices, and actions that all interrelate in well-defined ways. Distinctions between legitimate and illegitimate actions and the criteria for accepting explanations will be drawn from the fundamental texts of the religion and their interpretation by particular officials. In the domain of sports, the various objects, roles, and actions generate a viable, con-sistent activity, and legitimate actions and explanations are validated by the formal and informal rules and the interpretations of referees and judges. Even our experience in the physical domain is generated by us. We create distinctions in language, with which we recursively interact, such as "heavy" or "close" or their more precise scientific counterparts. However, Maturana does point out that the physical domain is different from other domains. It is the one in which we realize our existence as living systems and the one in which all others intersect through the operations of the nervous system and the body.

This description appears to be close to the ideas of Schutz (1967) in phenomenological sociology that individuals experience their lives within finite provinces of meaning and that consequently there exist multiple realities. However, Schutz's concept is tied to the individual, experiencing subject whereas Maturana's emphasis on the consensual nature of language places him nearer the intersubjective formulation of Wittgenstein's forms of life (Mingers, 1984).

7.4.4. The Domain of Science

One of these domains of reality, and thus of explanation, is science. This domain is particularly important because of the effects it has on modern society, and it is distinctive in being concerned essentially with explanation itself. Yet ultimately it is only another, equally valid, domain. Scientific activity and thought is no different in principle from other daily activity.

All domains of reality are associated with particular moods or emotions (see Section 6.3), and science is motivated by curiosity, a particular "desire for explaining." That is, scientists are particularly rigorous in ensuring that an explanation fulfills its criterion for the validation of explanations—that is, the scientific method. Indeed, this is how science gains its validity: not from describing or analyzing independent real-world entities, but by rigorously applying the scientific method to reformulate explanations in terms of the operational coherences that apply within particular domains of experience.

> Since the criterion of validation of scientific explanations is defined and constituted only in terms of the *operational coherences* of the domain of experiences of standard observers, it does not entail any supposition about an objective independent reality. . . . The claim that scientists make about the universal validity of scientific explanations [refers] to their validity through the application of the operational coherences that they entail in the world or worlds brought forth. . . .
> Maturana (1990, p. 20, my emphasis)

The basis for scientific validity is, therefore, a correspondence between the explanation and our experience; it is the operational coherences of our experience that are captured in the hypothesis. Note that it is not based on a correspondence between the explanation and external reality. Thus scientific explanations are valid because they are the application of operational coherences, but this is because, at the same time, "every system or mechanism operates only if the operational coherences that it entails are satisfied (Maturana, 1990, p. 32).

Maturana's early view of the scientific method was set out in Section 7.3.2.3, and in his more recent work it has been made more consistent with the overall theory. It still embodies the central idea that an explanation involves the positing of a mechanism that could generate the observed phenomena. The main development is that it is more recently formulated exclusively through experiences of the observer rather than events in the world:

> i)   The presentation of the experience (phenomenon) to be explained in terms of what a standard observer must do in his or her domain of experience (praxis of living) to experience it.
> ii) The reformulation of the experience . . . in the form of a generative mechanism that, if realized by a standard observer . . . would allow him or her . . . to have the experience to be explained.
> iii) The deduction from the operation of the generative mechanism proposed in ii), as well as from all the operational coherences of the domain of experiences of a standard observer entailed by it, of other experiences that a standard observer should have . . . and of the operations that he or she must realize in his or her domain of experiences to have them.
> iv) The experience by a standard observer of the experiences (or phenomena) deduced in iii) through his or her realization in his or her domain of experiences of the operations also deduced in iii).
> Maturana (1990, p. 18)

Several of Maturana's conclusions are worth highlighting. First, science inevitably occurs within a community of scientists or observers, based on communication and consensus about actions and their coordination. Within this consensual community scientists are essentially interchangeable and are likely to make similar observations and explanations (the standard observer). However, there can be several different communities, validating different distinctions and explanations. This view clearly mirrors Kuhn's work. Second, since the methodology only refers to experiences and activities of the observer, the resulting explanations cannot directly refer to or denote a world independent of the observer. Scientific validity therefore does not stem from correspondence to an independent world but from correct application of the scientific method and experience of the consequences of our actions:

> In fact, scientific explanations do not explain an independent world, they explain the experiences of the observer, and that is the world he or she lives.
> Maturana (1988b, p. 38)

7.5. Maturana's Philosophy—A Phenomenological Biologist?

In overall terms, however, Maturana's ideas show an amazing number of similarities to the phenomenology of Husserl and Heidegger, although he makes no reference to them. Winograd and Flores (1987a) have also juxtaposed Maturana and Heidegger in their important book *Understanding Computers and Cognition*. It is worthwhile demonstrating these concurrences in detail. First we must sketch the related but different concerns of Husserl and Heidegger.

7.5.1.   Husserl's Transcendental Subjectivity

Husserl's project was the reformulation of philosophy so that it could properly form a totally secure foundation for rational knowledge (Husserl, 1970, 1977; a good introduction is Hammond *et al.*, 1991). His method was to follow Descartes and doubt everything that was not absolutely certain in order to discover a foundation that could not be doubted. Such an ultimate foundation must be both *primary* (that is, not dependent on anything else) and *apodictic* (that is, absolutely indubitable). Does not the existence of the everyday world provide its own certainty? It certainly appears to be primary in that it seems to exist independently of us. Yet is this really apodictic? Husserl notes that our perceptions of the world can be mistaken, and that, in fact, we cannot actually distinguish between reality and a dream.

Husserl's response is not to reject immediately the reality of the world, but to put it to one side and suspend judgment about its existence. His way of doing this is to put the world "in brackets," to parenthesize it. This is called the "phenomenological *epoché*." As philosophers, we talk as if the world existed, but this is now something which must be shown. The world "claims being." If we bracket all that we normally take for granted, in what Husserl calls the "natural attitude," what then is left?

> If I put myself above all this life and refrain from doing any believing that takes "the" world straightforwardly as existing—if I direct my regard exclusively to this life itself, as consciousness of "the" world—I thereby acquire myself as the pure ego, with the pure stream of my cogitations. Husserl (1977, p. 21)

What is left is the self, the ego which is experiencing the world and reflecting on this process. In other words, despite suspending the existence of the whole of the world, there must still be some "thing" which is making that suspension. Moreover, Husserl argues that one now has access to the pure ego, uncontaminated with the assumptions of everyday life. This is the *transcendental ego*, and it is this which Husserl claims is the proper, indubitable foundation. It is more primary than the world because it remains after the *epoché*—it exists whether or not the world does—and it is more primary in the sense that it gives sense or meaning to the world. The world is experienced (by me) only through the possibilities allowed by my ego:

> I can enter no world other than the one that gets its sense and acceptance
> or status in and from me myself.
> Husserl (1977, p. 21)

The transcendental ego must also be apodictic since, in doubting everything there must still be some process, some subjectivity doing the doubting.

An alternative justification for this approach lies in the problem of trying to bridge the gap between the knower and the known—the naive realism of the natural attitude, which takes for granted the correspondence between our experiences and reality.

> But how can we be certain of the correspondence between cognition and
> the object cognized? . . . How do I, the cognizing subject, know if I can
> ever really know, that there exists not only my own mental processes,
> these acts of cognizing, but also that which they apprehend? How can I
> ever know that there is anything at all which could be set over against
> cognition as its object?
> Husserl (1964, pp. 15–16)

The phenomenological answer is to give up the natural attitude and its assumptions and instead to explore the experiencing subject in as uncommitted a way as possible. The method for doing this, the phenomenological method, is based on the *intentional* character of consciousness. This simply means that consciousness is always consciousness *of something*. We cannot simply be conscious; we must be thinking about something. By studying the objects of experience, not in the real world but in our experience, Husserl aims both to discover *a priori* rules or structures that govern our experience and to show that the ego displays the same structure. This is to be done by gaining pure descriptions of our experiences by disengaging the ego from the demands of the real world through the *epoché*.

First, one chooses a particular intentional object and then, after bracketing one's natural attitude toward it, begins to describe the actual appearances and the possible appearances that it could have in consciousness. One also describes the actual and possible ways that it can be experienced, e.g., the various perceptions one can have of it. This reveals both what one means or intends by a particular object and how one constitutes it. This method can equally be applied to things experienced physically, in which case it reveals the various dimensions of our spatial experience, and to more abstract terms such as "existence" or "truth." What do we mean when we use such terms? How do we constitute them? Husserl calls such theories "constitutional" theories.

The other step is a description of the subject, the ego itself. This is done using the same method, and Husserl identifies three key features: the subject has identity, the subject has individuality, and the subject's individuality is based on its experiences (of objects). The subject has identity because it is always the same subject experiencing, constituting, and synthesizing its life. The subject has individuality because it is different from other subjects; it has its own history of particular experiences leading it to have a particular character or style. Finally, it will have so developed through the particular objects it has experienced and might in the future experience, it will have developed a "habituality" of objects and types of objects.

This habituality, this set of readinesses to perceive (Vickers, 1968) and experience particular types of objects (e.g., physical ones), is not just a secondary effect, but is absolutely necessary for being a transcendental ego. It is this ability, developed through our history of experiences since birth, that enables us to synthesize the different sense experiences of a particular type of object and experience it as such an object. There must therefore be basic structures of the ego that govern our constitution of the world and must be the same as those governing the external world, for otherwise we could not have experiences of it.

Thus, although we have no access to an independent world, "objects exist for me, and are for me what they are, only as objects of actual and possible consciousness" (Husserl, 1977, p. 65), this does not matter since we can discover its structure by analyzing the experiencing subject that generates it for us.

> The attempt to conceive the universe of true being as something lying outside the universe of possible consciousness, possible knowledge, possible evidence, the two being related merely externally by a rigid law, is nonsensical.
> Husserl (1977, p. 84)

7.5.2. Heidegger's Phenomenology of the Everyday World

Heidegger, who was a student of Husserl's, shares the basic stance of phenomenology as directing its attention toward the experiencing subject, but he moves in the opposite direction. Instead of bracketing the natural attitude of the everyday world and moving toward purer and purer experience, he actually makes the everyday world of existence the focus of his enquiry. In his first work, *Being and Time* (English translation, 1962), he undertakes a phenomenological description of the way we ordinarily and routinely, indeed subconsciously, experience our world. Dreyfus (1991) provides an excellent introduction to Heidegger.

Humans are entities that can be self-conscious—that is, aware of their own selves—and, in particular, the way that the self experiences the world. It is this process of experiencing the world that is the way of being of humans. Much of the time we are not actually conscious of the way that our consciousness is working; we are concerned with the result of the activity that we are engaged in, and not what is happening in the mind. This is precisely what Heidegger is trying to describe or analyze: the way in which our self works, of which we are often not conscious or aware. He refers to this self, this entity that can inquire about itself as *Dasein* (literally, "being there").

He wishes to inquire into (that is, to make clear) the meaning of being, that is, the ways in which we experience and act on the world as entities that can actually reflect on our own selves. This must begin by looking not at scientific or philosophical consciousness but at the most basic everyday consciousness of existing and acting in the world, "Being-in-the-World." (Heidegger's language is full of compound capitalized expressions like this, which can be off-putting.) In so doing, what does the self *(Dasein)* come up against? First, all the objects and things in the world that are not selves; second, other selves; and third, its own self. Moreover, in relating to the world, in existing in the world, our basic attitude is always (except in pure contemplation) one of doing, acting, having some aim in mind, having some concern.

7.5.2.1. Our Experience of Objects

The world we meet is full of objects, and we can pick out and describe these objects—measure them, weigh them, analyze them, and so on. This is not, generally, how we relate to them. The way that we relate to or experience them is by using them in our direct activities. Indeed most of them we do not notice at all because they are not relevant to our immediate concerns. Generally, objects have been created or adapted precisely in order to be used, and they form part of a set of objects (Heidegger uses the term *equipment*), usually positioned with regard to their use. Given our particular purpose at any time, there will be a set of objects *ready-to-hand* (their mode of being) for use, and these will be selected by our consciousness as and when needed. Then, in use, they will actually disappear from our consciousness again as we concentrate on the job in hand. In typing on my word processor, I am concentrating on the meaning and am unaware of the keyboard. Generally, therefore, most of the world does not impinge on our consciousness at all; an object is either irrelevant to our task or is so immersed in the task that we are blind to it.

We become aware of an object in itself, it becomes an object for us, when something goes wrong with what we are doing and forces us to pay attention to the tools themselves. Something is broken or unusable or is missing or prevents further progress. Then the tool, its uses, composition, and surroundings, become an object of our consciousness. We now look at them in a different way in order to solve our problems. The being of such objects is *present-at-hand*. Just as our usual attitude toward objects is one of use rather than of analysis or description, so our everyday experience of space is through our activity rather than objective measurement. We experience space through objects and their relations to one another—"next to," "on top of," "beside," and so on. Objects exist "in their place," not at some particular co-ordinate position. Distances and times are thought of generally and with regard to our activities, not in absolute terms—"a good morning's drive." Objects are close or far in how easily they can be used—do they have to be collected? Fixed? Found? Or are they close by and ready? *Dasein* is characterized by a general attitude toward the world of objects, that of concern, as in "to be concerned" that something is going well.

7.5.2.2. Our Experience of Other People

In our normal daily activities, we encounter others and relate to them according to their place within the structure of the activity—that is, as customer, seller, deliverer, lecturer. This is not to say that we thereby treat them the same as objects; rather, we recognize them as other *Daseins*. Thus part of our Being-in-the-World is *Being-with* others. This is part of our nature even if we happen to be by ourselves. We are generally conscious of only those others whom we actually deal with—not all others in general—and we may well be physically present with others (for example, in a public place) and yet not be interacting with them at all. They may just be present-at-hand. Heidegger terms our basic attitude to these others *solicitude*—suggesting the idea of caring for or concern for a person—or its lack.

7.5.2.3. Our Everyday Experience of Ourselves

Generally, we are so immersed in our activities that we do not consciously reflect on our Being. Instead, being involved with other people, we are constantly comparing ourselves with them and being concerned about the extent to which we differ. Our concern not to differ leads us not to be ourselves, but to be subjected to people in general—"they" the crowd, the mass. We take our opinions and beliefs from them (through the media). Their way of existing is to emphasize the average and maintain averageness. So, in being involved in the everyday world, we do not live our own lives for ourselves, but live lives determined by others. Nor are we aware of this. Our attitude toward our own self-knowledge is characterized by *transparency* or its lack.

7.5.2.4. The Process of Being-in the Everyday World

We have looked at consciousness of the world and its objects, of other people, and of ourselves. Now we look at the actual process of being conscious. This can be investigated by recognizing that we are always at a particular place and time, being active and trying to bring something about. We are always "here," but our consciousness is trying to get "there" *(Da-sein)*. It is not closed off, but open to the world through its trying to bring something about. In so doing, consciousness is characterized by its state of mind, or mood (the state it is in) and by its understanding (its projecting of its possibilities in the future). These are both expressed (and experienced) in discourse or language.

We are always in some state of mind or mood. We cannot "not have" a state of mind. Our mood reveals, and is a result of, how we have been doing. We must already be in a mood before we can be consciously aware of what it is. We find ourselves "thrown in" to a particular situation with a particular state of mind *(thrownness)*. We may go along with it, or exchange it for another, but we cannot be without one. States of mind reveal our consciousness as thrown-in, as predisposed in a particular way. They are prior to thought and willpower. Our mood will have already shown or affected (disclosed) our Being-in-the-World as a whole. What we experience is experienced through our mood. Our state of mind controls what we can be conscious of, what can matter to us or be of concern. Things cannot cause us fear or anger or joy unless our state of mind is such as to allow this. We become afraid of something before we are aware of being afraid, and by the time we are aware, we are already in a state of mind of fear.

Our mood is the state we are in; our understanding reveals our possibilities for action. In being there, in the world, we point to the fact that there is a world, of things, others, and self, toward which we act. At any instant, there are a number of possible ways that we could act— a number of ways that we could be. It is our understanding that reveals these possibilities to us. These possibilities are not determined outside of ourselves, but through our state of mind. They continually occur and are rejected or accepted. So the world is always a world of possibility, not of actuality—possible ways of being revealed by our understanding that what we are is what we become (i.e., which of the possibilities occurs) and this then affects the state of mind, which throws up the next possibilities, and so on.

Having become conscious of something through understanding, we may decide to make it more explicit, to take it apart, to lay it out and see how it fits together. We will exemplify this with the world of objects, but it applies equally well to the worlds of people and of the self. Whenever we interpret (or analyze) something, it is always from a point of view; given the way that we were going to use it, we always see it as something. Also, the object is not isolated, it is always part of a whole complex of things that go together for a particular purpose. We already have some understanding of what this structure is, although it may be far from perfect, and of our immediate context. We have some knowledge of its structure as something. When we have produced an interpretation of something it then has meaning for us. Meaning is always created by the *Dasein* in its interpretation of the world. Things do not have meaning in themselves. So to ask for the meaning of something is to set it out and explain it from the point of view of a conscious self active in the world.

7.5.2.5. Discourse and Language

Our states of mind and our understanding are intelligible to us because they can be articulated or expressed in speech as discourse. Discourse is communication, but not objective communication from one fixed being to another. Rather, it is part of the process of creating shared states of mind:

> Communication is never anything like a conveying of experiences, such as opinions or wishes, from the interior of one subject into the interior of another. Dasein-with [the process of being with others] is already essentially manifest in a co-state-of-mind and a co-understanding. In discourse Being-with becomes "explicitly" shared; that is to say, it is already, but it is unshared as something that has not been taken hold of and appropriated.
> Heidegger (1962, p. 205)

This has been only a brief sketch of the very early part of Heidegger's philosophy, but it is sufficient for our purpose.

7.5.3. *Maturana as Phenomenologist*

The most important similarity among all three writers is the recognition of the problematic nature of external reality and the decision to turn inward, to examine and analyze the nature of the observer, the person who experiences and gives meaning to the world. For Heidegger and Husserl this is at the level of mind and consciousness, while for Maturana it is based on a theory of the biology of the cognizing system. This general insight is shared by other schools of thought, but there are also many detailed parallels, which are listed in Tables 7.1 and 7.2. However, Husserl and Heidegger both move in very different directions and it is necessary to summarize this debate and locate Maturana within it.

Husserl and Heidegger are both known as phenomenologists, and yet they focus in diametrically opposed directions (Boelen, 1975). A good way to locate this difference is to consider Descartes' famous *cogito ergo sum*—a starting point for both (F. Smith, 1970). We can then say that while Husserl was concerned with *cogito* (thought and consciousness) Heidegger was concerned with *sum* (being and activity).

For Husserl, there is a road to true knowledge through disinterested contemplation. Phenomenologists should detach themselves from everyday concerns in order to gain access to the pure ideas of the ego's consciousness. For Heidegger, there can be no such idealist representations of subject and object. Our natural attitude, our being-in-the-world, cannot be expressed in, nor does it consist of, conscious beliefs, ideas, rules, and intentions. Rather it is a sub- or preconscious attitude socialized into us and embodied in our actions and skills. Being, our way of interpreting and dealing with the world, is inherent in the practices of our culture and society and is continually enacted by us in an unmindful way. We cannot uncover the beliefs or intentions behind what we do, for there are none; there are only skills and practices.

TABLE 7.1
Similarities between Maturana and Husserl

1. Both make the point that our perception of the world is corrigible, and indeed that we cannot distinguish between illusion and reality.
2. Maturana follows a course similar to Husserl's and puts objectivity (rather than the objective world) in brackets in order to see what remains when one does.
3. Both emphasize how we are unavoidably constrained to experience our own, individual, experiential world and that we can never compare this with some other, objective world.
4. They both use the term *constitutive* to connote the bringing into being of our reality.
5. In describing the subject, both pick out identity and individuality as key features. This indeed is the starting point for the theory of autopoiesis.
6. Husserl's concept of the habituality of an individual is very similar to Maturana's structural coupling. An individual, through its history of interactions, develops a particular structure in relation to its environment, which determines the interactions that it can have.

It seems clear that in this debate, Maturana is in Heidegger's camp. He, too, emphasizes that cognition is not detached cogitation but situated, practical action. This is brought out even more in Chapter 11, where Varela's recent contribution of a theory of embodied cognition draws on another phenomenologist, Merleau-Ponty, whose work develops further the road taken by Heidegger away from pure consciousness.

To conclude this section, I believe the similarities outlined above do justify the claim that Maturana is best seen as a phenomenological biologist. However, there are a couple of points still to be made. First, in aligning Maturana with such major philosophers I am not thereby trying to validate his work. Nor, indeed, am I trying to justify phenomenology. As will be demonstrated in Section 7.6, I believe there are major problems with this position. What I am trying to do is to situate Maturana's work within conventional academic disciplines both to show that it is

TABLE 7.2

Similarities between Maturana and Heidegger

1. The starting point for both is the involvement of people in their practical daily life—concernful activity. This is the basis for language, and other cognitive domains such as science are essentially derivative from it and similar in nature to it. Maturana stresses how language itself emerges through the need for the coordination of our practical activities.
2. Heidegger's idea of thrownness is also present in Maturana; see, for example, Maturana and Varela (1977, p. 242), where he says that we always have blind spots, things that we do not see, which we become aware of only when we are shocked or dislodged from our normal attitude.
3. Both stress the importance of emotion or mood as an ever-present background which conditions how we experience the world. Maturana's triad of language, emotion, and body fits well with Heidegger's description of mood, understanding, and practical (physical) activity.
4. Both discuss how even the seemingly most objective domain, the physical world, is constituted for us by the practical distinctions that we make in everyday life.
5. Both identify a particular, quite similar, attitude toward other people as constitutive of the social domain—solicitude, or care for another (Heidegger), and mutual acceptance (Maturana, 1988b, although in 1980b, p. 15, he also uses the term "care").
6. Maturana's idea of operational coherences seems very similar to Heidegger's analysis of understanding, in which he claims that we always interpret things with some degree of preunderstanding and always see them as part of a complex or system of objects and processes that go together.
7. Both recognize that communication between people is not some objective exchange of symbols but relies on an already existing similarity of structure between the communicators—structural coupling.
8. Finally, as the quote at the start of Chapter 1 shows, Heidegger all but produced the terms "autopoietic" and "allopoietic" himself.

not as singular or radical as might at first appear and to stimulate interchange of ideas among different domains.

Second, I think that in a particular respect Maturana's work represents a distinct advance on classical phenomenology, a major criticism of which is that it is essentially individualist and has great difficulty in generating the intersubjective nature of social reality (Mingers, 1984, 1992a). Here, Maturana begins from an intersubjective position. We are (as self-conscious beings) constituted through our language, and language is inevitably an intersubjective phenomenon. As Wittgenstein (1978) also argued, there can be no such thing as a private language. Thus language is essentially a consensual domain of agreements, of structural coupling that permits the operations of observers.

The work of Merleau-Ponty (1962, 1963), based on both phenomenology and the natural sciences such as physiology and psychology, is very relevant and will be discussed in Chapter 11.

7.6. A Realist Critique of Maturana's Constructivism

From what has been said, it would appear that Maturana must be decidedly antirealist. However, the heart of his methodology corresponds exactly with that of the critical realist view outlined in Section 7.2, namely, that science should proceed by hypothesizing mechanisms or structures that, if they existed, would generate the phenomena to be explained. The central difference is the ontological status of such entities. Realists argue that, putatively, they exist in a world independent (at least in the natural sciences) of the beliefs of the observer. Maturana would hold that they are constructs of the observer, as a member of a community of observers, which do not represent or denote anything independent. This difference is reflected in methodological terms. The realist attempts to demonstrate the existence of mechanisms retrospectively, while Maturana predicts the occurrence of other phenomena of experience.

In early papers, Maturana conceded the need to assume some substratum of reality, but more recently he argues repeatedly that there is no independent reality, not merely that we have no access to such a reality:

> . . . the physical domain of existence is one of many domains of reality that we bring forth as we explain our praxis of living. . . . outside language nothing (no thing) exists because existence is bound to our distinctions in language. . . . I am saying that all phenomena . . . are cognitive phenomena that arise in observing as the observer operates in language. . . . Nothing precedes its distinction; existence in any domain, even the existence of the observer themselves, is constituted in the distinctions of the observer.

> . . . if we ask for the characteristics of the transcendental substratum on
> which, for epistemological reasons we expect everything to take place, we
> find . . . that we cannot say anything about it, not even to refer to it as an
> it, because as soon as we do so we are in language.
> Maturana (1988b, pp. 79–80, my emphasis)

What are we to make of this? Are we really to believe that while Maturana writes an article he believes that he is bringing forth the pencil and paper that he uses in just the same way as he brings forth the ideas that he writes down? To believe that we literally create the world we describe takes us toward the realm of Berkeley's "to be is to be seen" but without the all-seeing God, a point also made by Zolo (1992).

Part of the answer is that Maturana uses the words *existence* and *to exist* in a rather unusual way (Maturana, 1991a, p. 389). From the point of view of objectivity-in-parenthesis (his point of view), our biological limitations as observers mean that we can never refer to or describe something independent of ourselves. But when we do distinguish an entity in our language or action, it can then be treated as if it were independent. This, for Maturana, is the only valid form of existence: "I indicate in the explanatory path of objectivity-in-parenthesis that existence is constituted by what the observer does, and existence refers to the conditions of constitution of what we talk about" (Maturana, 1991a, p. 389). Since we can never get out of language, we can never use "existence" to connote an independently occurring entity. For us, all that can exist is what we constitute.

Maturana also makes it clear that he does not (and could not) seek to explain "reality" (Maturana, 1991a, p. 386). Rather, what he wishes to explain is our experiences since that is all we have access to:

> . . . every explanation is given by an observer as the proposition of a gen-
> erative mechanism that uses underline experiences to generate underline experiences within
> the regularities of underline experiences as an answer to a question that accepts an
> explanation as an answer.
> Maturana (1991a, p. 390, my emphasis)

Thus, for Maturana, we remain in a closed domain. We begin with experiences that we wish to explain, we propose explanations of the operational coherences of our experience, and we make predictions of other experiences that we or others may have. We can never, in any way, escape from experience and language to reach the shores of pure reality.

This is really the heart of Maturana's radical position, and I will argue that his position is ultimately inconsistent. I accept that his ideas lead inescapably to the view that we cannot directly access a world independent of our perception and language. This does not, however, prove that there is no such world, as the arguments outlined below show (Mingers, 1990).

As Luhmann, who has developed a theory of autopoietic society (see Section 8.5), says:

> If a knowing system has no entry to its external world it can be denied that such an external world exists. But we can just as well—and more believably—claim that the external world is as it is. Neither claim can be proved; there is no way of deciding between them.
> Luhmann (1990b, p. 67)

Similar criticisms of Maturana's constructivism have been put forward independently of mine by Zolo (1992), Held and Pols (1985a, 1985b, 1987a, 1987b), and Johnson (1989, 1991a, 1991b, 1991c, 1992, 1993a, 1993b). Held and Pols will be discussed further in Section 10.4, as a similar debate has occurred within family therapy.

7.6.1. The Constraints of Reality

First, if there were no reality independent of our descriptions, then we would be free to bring into existence any world that we desired. In fact, reality constrains the success of our ideas and theories. While we are free to imagine what we wish, doing so does not make it the case. A belief in the ability to fly or stay under water for ten minutes will be refuted by death itself. This is essentially the position of von Glasersfeld, a writer with ideas similar to those of Maturana, who describes himself as a radical constructivist. He eschews the idea of correspondence to an independent reality but accepts that reality limits what is possible and that knowledge may therefore fit or not fit such a reality (Von Glasersfeld, 1984; Kenny, 1988).

7.6.2. Contradictions within the Theory

The next point to make is that Maturana's theories, when taken together with the claims that he makes from them, are self-contradictory. They are, in fact, inconsistent on two different levels. The first is common to all strongly relativist theories. Such positions make an epistemological claim that all knowledge is relative to the knower (or community of knowers), that is, that no theory can claim objective truth. However, since such a theory is self-referential, it must equally apply to itself. If this is accepted, then we are not compelled to agree with the theory; if it is rejected, then the theory is not consistent. This problem is clearly exemplified in Maturana's work. On the basis of a study of the biology of the observer, he claims that we have no access to independent reality and that different explanatory domains are equally valid. He must either accept that his whole theory has no special claim to validity or exempt his own particular theory from his stipulations, which would be inconsistent. As Held and Pols write:

> Of course, Maturana *is* making a reality claim—a general claim about the nature of the observer or knower; and *that* is what generates the contradiction, for he also claims that each cognitive act produces its own subject-dependent reality.
> Held and Pols (1987b, p. 467, original emphasis)

The second manifestation of the contradiction is that Maturana's own theories specifically require that there be an independently existing world. The notion of something outside the individual subject occurs in various guises—the autopoietic organism exists within a *medium* which supplies basic chemicals necessary for continued autopoiesis; the observer sees an *organism* within an *environment* and the organism itself occupies a *niche*, that is, a domain of possible interactions; changes within a structure-determined system are triggered or selected by a deforming agent or by other systems; autopoietic systems become *structurally coupled* to their medium and to other systems; language, as a consensual domain coordinating action, requires *more than one observer;* and science depends on a *community of scientists.*

Maturana could, and probably would, argue that these are all distinctions constituted by himself, the observer, and make no commitments to an independent ontological existence. Against this I would use the original Cartesian argument as refined by Husserl. Descartes' method was to doubt the existence of everything but to conclude that there must, in the end, be *someone* doing the doubting. Husserl argued that it was mistaken to infer the existence of an actual person, be it a mental or physical entity, but that one could infer the existence of some process or subjectivity making possible the doubt. Equally with Maturana, we may deny reality to all the autopoietic constructs, but ultimately there must be some observing process generating them, or there would be nothing.

The third point is that, in describing autopoietic unities, Maturana and Varela state that they produce their own boundaries, which the observer may mistake:

> Since it is a defining feature of an autopoietic system that it should specify its own boundaries, a proper recognition of an autopoietic system as a unity requires that the observer performs an operation of distinction that defines the limits of the system in the same domain in which it specifies them through its autopoiesis. If this is not the case, he does not observe the autopoietic system as a unity.
> Maturana and Varela (1980, p. 109).

In all these ways, Maturana's ideas rely on a reality independent of an individual and therefore equally independent of a group of individuals.

7.6.3. *The Necessary Preconditions for Science*

There are philosophical arguments, largely developed by Bhaskar (1978, 1979), supporting critical realism. The main thrust is to ask what the world must be like for science, as we know it, to exist and be intelligible. What is presupposed by the activity of science and is thus transcendental? The answer is a world of entities independent of our descriptions of them. These Bhaskar calls the "intransitive" objects of knowledge, in contrast to the experiences, theories, and descriptions used in the production of knowledge (the "transitive" objects).

We can imagine the world existing without human beings to observe it, and therefore with no science to describe it. Our knowledge suggests that it has been so for most of time. However, is it possible to imagine us observers existing without a world in which to exist? For science to occur with no intransitive objects?

> It is not necessary that science occurs. But given that it does, it is necessary that the world is a certain way.
> Bhaskar (1978, p. 29)

All views of science accept that it is based in experience, although they may differ about the nature and cause of experience. Our scientific experience occurs through scientific activity (the praxis of living, for Maturana) and consists of both perception and active intervention through experimentation. Moreover, science has a history of change; theories develop and in time are replaced by quite different theories. All these characteristics of science are difficult to sustain from a constructivist position.

First, in the case of perception realists argue that since it is possible to have different experiences of an object—to perceive it differently (e.g., visual illusions)—and since it is necessary to train people to perceive scientific data correctly (e.g., cloud chambers or X-rays), then these objects must be independent of our perceptions. This is a neat reversal of the subjectivist argument that our perceptions cannot be trusted. Maturana often argues that the nature of the nervous system means that it, and therefore we, cannot distinguish the origin of neuronal activity, and thus cannot distinguish reality from hallucination. However, he does agree that the observer can so do by observing both the organism and its environment. Moreover, since we have the possibility of being self-observers, we must at least sometimes be able to do so ourselves (for example, to realize that the experience of an amputated leg is not real).

Second, concerning scientific experimentation, Bhaskar argues that in an experiment we deliberately bring about a sequence of events that would not otherwise have occurred, at least at that point in time. In doing so, we cause the sequence of events but not the laws that they obey. Moreover, we find that these laws apply in situations other than those of the experiment, as well as in other similarly contrived situations. This is also assumed within Maturana's methodology. All of this implies that the causal laws operate independently of the observer.

Third, constructivism finds it difficult to explain scientific change and, *a fortiori*, scientific progress. If there is no external reality that constrains our theories or to which our theories might refer, then why should we change them and why should we prefer one to another? Indeed, are we even in a position to compare descriptions that, having nothing to refer to, have nothing in common? As Bhaskar has argued (1986, p. 72), accepting that beliefs and descriptions are historically and culturally conditioned (epistemic relativism) does not force us to accept that all beliefs are equally valid (judgmental relativism). For Maturana, descriptions are characterized as scientific if they are generated following the scientific method (as described above). However, this by no means overcomes the difficulties. First, it does not avoid the necessity for choice since competing explanations are produced, yet the method does not tell us how to make such a choice. Second, it too implicitly rests on an assumption of an independent world. Even though it refers only to subjects' experiences, the movement from Step ii (hypothesizing a mechanism) to Step iii (predicting novel experiences) is intelligible only on the assumption of stable, enduring causal structures that have the same effects at other times and on other observers.

7.6.4.   The Epistemic Fallacy

The second argument concerns the relation between ontology (questions concerning what exists) and epistemology (questions concerning what we know). In denying the existence of independent entities on the basis of theories about the observer, Maturana is reducing ontological questions to epistemological ones (the epistemic fallacy), that is, trying to answer questions about what exists purely in terms of our knowledge of or about what exists. While it is true that our knowledge limits what we can know to exist, it does not follow that it can limit what actually does exist. The causality, for at least some objects, must be the other way around—observers can know because they exist; they do not exist because they know. To thus reduce ontology to epistemology is to mistakenly make human beings and their experiences the measure of all things.

7.7.   Conclusions

It has been the argument of this chapter that Maturana's espoused position is ultimately inconsistent but that it can be successfully reconstructed in the light of critical realism as follows: There is a single, real, materially existing world. This has, through processes of evolution, generated organisms capable of creating distinctions, descriptions, and constructs, subject only to their own internal structure. This leads, in such organisms, to a proliferation of domains of experience and interaction essentially free from dependence on and determination by the material world. One of these domains is that of science, in which observers cast their net of descriptions back onto the world itself. These are free human constructs, based always on subject-dependent experiences, yet the world that they relate to is independent of the observer's descriptions and existed prior to them. To the extent that the mechanisms hypothesized in scientific descriptions do exist in the material world, then the praxis of observers based on these descriptions is more successful. The contingent fact that science has been successful in enlarging our domain of interactions is, if not proof, at least strong evidence for such existence. What can, in any case, be shown is that the fact that our descriptions are always subject-dependent does not preclude the existence of a world independent of such descriptions.

III

Applications Of Autopoiesis

Autopoietic Organizations and Social Systems

It may be that "linguistically generated intersubjectivity" and "self-referentially closed system" are now the catchwords for a controversy that will take the place of the discredited mind–body problematic.
Habermas (1990, p. 385). This is the final sentence of the book, written in relation to Luhmann's autopoietically based systems theory.

8.1.   Introduction

The concept of autopoiesis clearly was developed in order to explain the specific domain of physical, living systems. However, from the start it has been suggested that other types of entities, particularly human Organizations ("Organization" will refer to clubs, businesses, etc., while "organization" will mean Maturana's and Varela's term) and societies exhibit the same characteristics that autopoiesis explains in physical, living systems; namely, autonomy and the persistence and maintenance of identity despite wholesale changes of structure and turnover of components. Therefore, might they too be autopoietic? Various authors have considered this question, and Stafford Beer was characteristically enthusiastic in his Preface to "Autopoietic Systems" (1975, p. 70):

> I ask for permission actively to enter this arena of discussion. . . . For I am quite sure of the answer: yes, human societies are biological systems. . . . any cohesive social institution is an autopoietic system—because it survives, because its methods of survival answer the autopoietic criteria, and because it may well change its entire appearance and its apparent purpose in the process. As examples I list: firms and industries, schools and universities, clinics and hospitals, professional bodies, departments of state and whole countries.

This chapter will consider the attractions and development of such ideas, pointing out the serious difficulties involved in the transfer of such a physically oriented concept to essentially nonphysical domains. A number of authors will be considered in detail, but others who have also discussed social autopoiesis are Benseler *et al.* (1980); Zeleny (1980); Roth *et al.* (1981); Ulrich and Probst (1984); Baert and De Schampheleire (1987); Brown (1988); Platt (1989); Burghgraeve (1992); Meynen (1992); and Bailey (1994). Espejo (1993) links autopoiesis to Beer's Viable Systems Model in analyzing organizations, and Leydesdorff (1993) has an interesting discussion of a relationship with parallel distributed processing systems—neural networks—which are explained in Chapter 11.

The central problem is that the autopoietic definition specifies the production of the components constituting the entity and the production of a boundary separating the entity from its environment. The definition does not specify that these must be physical components, but if they are not, then what precisely is their domain of existence? This chapter is structured by different responses to this central question.

The first group of authors, in one way or another, do not really recognize that the problem exists but, in my view, simply apply autopoiesis naively to the social domain. This includes Beer, Zeleny, and Robb. The second response is to accept that social systems are not autopoietic as such. Varela defines a closed but not autopoietic organization, while Maturana argues that they are simply the medium in which autopoietic systems interact. The third response is to modify or enlarge the definition of autopoiesis. Luhmann conceptualizes the nonphysical production of events—society as autopoietic communication. Finally, autopoiesis can be used simply as a metaphor, generating interesting ways of seeing social systems without requiring the ontological commitment that they be autopoietic. As an example of this view I discuss Morgan's work.

8.2. The Simplistic View of Organizations and Societies as Autopoietic

8.2.1. Autopoietic Organizations

Beer has pointed to the main attractions of applying autopoiesis to Organizations and societies in the quotation above. Many social institutions, from small clubs, groups, and families, through varying sizes of Organization, right up to societies, countries, and cultures, exhibit a tremendous, longterm stability and persistence. Despite significant changes in their environment and tremendous internal structural changes of both members and relationships, some such entities have maintained a continual identity over long periods of time. In many cases (e.g., some religions and culture) this is in the face of deliberate and sustained attempts to destroy them. Are these not precisely the characteristics that the idea of a self-producing system can explain?

Beer sees autopoiesis as complementary to his own substantial, biologically based, work on viable systems (1979, 1981). Beer's theory describes what he considers to be the necessary organization for any system (biological or social) to be viable. The basic model consists of a recursive nesting of five subsystems. System 1 consists of the basic productive units of the Organization, each of which is autonomous. System 2 coordinates the operations of the units of System 1. System 3 is the central internal control mechanism for the Organization. System 4 is the intelligence function, scanning the environment and planning for the future, and System 5 gives overall purpose and policy to the Organization.

Beer argues both that the Organization as a whole is autopoietic and that the units of System 1 are autopoietic, but that Systems 2 to 5 should not be. He considers (like Robb, see below) that autopoiesis can give rise to pathology in Organizations. This can occur in two ways: the Organization as a whole may lose sight of its overall purposes and objectives, and its primary activity may become its own self-production (i.e., it is simply "going through the motions"). Or the systems which should not be autopoietic (2–5) may become so. Then particular parts of the Organization become self-serving and self-producing at the expense of overall viability.

As well as Beer, other authors easily made the assumption that social systems could be seen as autopoietic and that human Organizations either were or should be designed to be. Facheux and Makridakis (1979), looking at the design of Organizations, contrast what they call an I/O control model with an autopoietic model:

> . . . autopoiesis is a characteristic and consequence of autonomy and self-reference. It is a process of creating oneself. Varela's views are valid for all living systems but take particular significance for social systems. . . . In the rest of this section, several other examples will be taken from four levels of organizational complexity which illustrate further the concept of autonomy and the antinomy between autopoietic and I/O type systems. Faucheux and Makridakis (1979, p. 216)

In a similar vein, Zeleny and Pierre espouse the idea that Organizations should be designed to be autopoietic:

> . . . managers as catalysts induce the components to make their own decisions, conduct their own analyses, select their own criteria. A unique

autopoietic organization, a network of values, norms and precepts, is self-
created, self-maintained and self-grown.
Zeleny and Pierre (1976, p. 163)

The main idea behind these views is that humans are autopoietic
entities and as such autonomous and independent. Traditional types of
Organizations, however, treat them purely as components of the system; that is, they treat them as allopoietic. Not only is this wrong in a
moral sense, but it is also not necessarily good systems design. Autopoiesis shows how systems can function in a decentralized, nonhier-
archical way purely through the individual interactions of neighboring
components.

Robb, in a series of papers (1985, 1989a–e, 1991, 1992a,b), has argued the case that there are what he terms "supra-human" systems that
are autopoietic and that this has serious implications for mankind, as
they are essentially out of our control. In early papers, he suggested
rather simplistically that the components of autopoietic social systems
could be taken to be "mind-sets," which consisted of "specific goals,
explicit statements of objectives, ideas about organizational structure (in
the operational sense), culture, and so on, and as actions taken in the
organizational context" (Robb 1989b, p. 345). He adds that boundaries
could be defined that would separate different actors in the situation
and even different thoughts within an individual. This characterization
of social autopoiesis has been criticized for the reasons to be discussed
in Section 8.2.2, suggesting that Robb's formulation "is much too vague
and contentious to stand as a *proof* that there can be autopoietic social
systems" (Mingers 1989b, p. 350); furthermore, it is suggested that Robb
specify in detail the components, processes, and boundaries of an example of a social autopoietic system.

Robb responded in a paper entitled "Accounting—A Virtual Autopoietic System?" (Robb, 1991) by trying to demonstrate that accounting
does form such a suprahuman autopoietic system. His first step is to
argue that there is a conceptual problem in recognizing autopoietic social systems since these are necessarily wider than ourselves, the observers. We can be only components of such systems and so cannot
interact with them as unities and specify their boundaries. This leads
Robb to suggest that at most we can characterize "virtual" autopoietic
systems, that is, systems that we think behave "as if" they were autopoietic systems. We should do this by trying to detect particular features
that might be associated with autopoiesis, such as mutual causal feedback loops. I have some sympathy with this idea, as it does reflect a
difference between social systems and physical systems, and it is actually quite close to Maturana's own scientific method (Section 7.3.2),

namely, the idea of hypothesizing mechanisms that, if they existed, would generate the experienced phenomena. However, simply looking for feedback processes and clusters of interacting components does seem rather weak.

Robb goes on to be more specific about the nature of the components of social systems, agreeing that his previous notion of "mind-sets" was "admittedly vague" (1991, p. 219). He now develops a view, based on Luhmann (1986) and Pask (1981) (and Maturana, although Robb does not actually say so), of conversations in which meanings emerge from linguistic interaction, are picked up and used in communications, and then give rise to new conversations and meanings. Thus, "autopoietic social organization is sustained by the *continuous* production and reproduction of meanings through communication and conversation, and *vice versa*, in the cognitive domain" (Robb, 1991, p. 220, original emphasis).

This certainly seems a better basis for the ascription of some kind of autopoiesis to social systems than did the earlier, fuzzy ideas of mind-sets, and the detailed description of the domain of accounting is quite interesting. However, as it is based on Luhmann's work, which is discussed critically and in detail in Section 8.5, and as there is also a chapter below on the law as autopoietic, I shall not undertake a thorough analysis here.

8.2.2. Difficulties of Social Autopoiesis

While the idea of autopoietic Organizations and social systems is very attractive, fundamental difficulties are involved in such an application. If the concept is only being used metaphorically, in order to help our thinking, then no great problems emerge—it is simply a matter of whether or not it is fruitful. To go beyond analogy, however, and claim that an Organization or a society is autopoietic, is to raise contentious ontological claims that in many ways lie at the heart of social theory and its debates between objectivism and subjectivism (Mingers, 1984): namely, to what extent can the terms that we use in social description (e.g., "middle class," "Organization," "Warwick University") denote objectively existing entities as opposed to being constructs of the observer? This is already explicitly addressed in autopoiesis at the physical level, where a clear distinction is drawn between the observer's descriptions and the operational autopoietic system. The problem is more acute, however, at the social level. Without discussing this difficult philosophical problem in general, a number of aspects particular to autopoiesis will be mentioned.

If the attribution of autopoiesis to social systems is to be more than
a woolly generalization, then we must examine carefully its specific definition. There are three essential elements:

1. Centrally, autopoiesis is concerned with processes of production—the production of those components that themselves constitute the system.
2. It is constituted in temporal and spatial relations, and the components involved must create a boundary defining the entity as
   a unity—that is, a whole interacting with its environment.
3. The concept of the autopoietic organization specifies nothing beyond self-production. It does not specify particular structural
   properties and thus shouldn't need to be modified to deal with
   social systems.

In applying these ideas strictly, there are obvious problems. Is it
right to characterize social institutions as essentially processes of production and, if it is, what exactly is it that they are producing? If human
beings are taken as the components of social systems, then it is clear
that they are not produced by such systems but by other physical, biological processes. If we do not take humans as components, then what
are the components of social systems? The emphasis on physical space
and a self-defined boundary is also problematic. While space is a dimension of social interaction, it does not seem possible to sustain the
central idea of a boundary between those components that are both produced by and participate in production, on the one hand, and those that
are not, on the other. Generally, people can choose to belong or not
belong to particular institutions and are members of many at any time.
What is it that would constitute the boundaries of such systems and,
moreover, how can it be said that such institutions act as unities—is it
not only individual people who act?

Overall, it seems difficult to sustain the idea that social systems are
autopoietic, at least in strict accordance with the formal definition. To
illustrate this more specifically, a detailed critique of a recent paper by
Zeleny and Hufford (1992a) will be presented in Section 8.2.3. In this
paper they claim not only that social systems such as the family are
autopoietic, but, more startlingly, that all autopoietic systems are social
systems. My critique is based on Mingers (1992b).

However, it is possible that the concept can be useful metaphorically in helping our thinking, or that a more generalized version, such
as Varela's idea of organizational closure (Section 8.3), could be fruitfully
applied. A more radical approach is to apply autopoiesis not to physical

systems but to concepts or ideas. Maturana defines a unity as "an entity, concrete or conceptual, defined by an operation of distinction" (Maturana, 1975b) and thus opens the possibility of an autopoietic conceptual system. Such a system might consist of ideas, descriptions, or messages that interact and self-produce. This approach could be related to concepts of other writers, such as Bateson's ecology of ideas (1973), Popper's World Three (1972), or Pask's conversation theory (1976). Luhmann's development of differentiated autopoietic society, which specifies communication as the basic component, will be discussed in Section 8.5.

8.2.3.   A Critique of Zeleny and Hufford's Social Autopoiesis

Zeleny and Hufford's paper aims to demonstrate that systems in three different domains are autopoietic—the eukaryotic cell (biological), osmotic growths (chemical), and families (social). The specific arguments of the paper are that the family, as an example of a "natural social system," is autopoietic; that all natural social systems are autopoietic; and, finally, that all autopoietic systems are social. I shall deal with the last contention first as it is the least sustainable even if it were the case that the first two were correct.

Zeleny and Hufford conjecture that "autopoietic systems, both 'organic' and possibly 'inorganic,' are necessarily social" (1992a, p. 156). Thus, by their own earlier demonstrations in the paper, not only are (biological) cells social, but so too are (chemical) osmotic growths. What are we to make of this transference of the social to the realms of physics and chemistry? Is it not so great a distortion of the underlying idea of "social" that the term becomes meaningless?

I would argue that there are at least two elements that must be common to all definitions of *social:* first, it relates to the activity of groups of entities rather than single individuals; second, it concerns rule-based behavior rather than physical cause and effect. Thus the behavior of billiard balls on a table could not in any sense be termed social. The authors would appear to agree since they say, "Components and participants in autopoiesis must *follow rules, interact and communicate*—they form a community of components, a society, a social system" (p. 157, my emphasis). While physical systems do consist of groups of components, it cannot be said that they follow rules. The essence of rule-governed as opposed to cause-and-effect behavior is that a rule can be broken, or followed in right or wrong ways (Winch, 1958; Wittgenstein, 1978). This is clearly not the case for physico-chemical interactions—

molecules cannot choose to interact or not: their behavior is determined. Indeed, Zeleny and Hufford implicitly recognize this when they refer to "social systems *proper* (i.e., human systems)" (p. 157, my emphasis). Thus to extend the term *social* to all autopoietic systems is to lose its essential meaning.

Zeleny and Hufford use the six-point key of Varela *et al.* (see Section 2.2.1) to identify autopoietic systems. It is not clear that this heuristic device actually captures the full richness of autopoiesis, but I shall restrict myself to criticizing Zeleny and Hufford's application of it. It does, in any case, clearly embody the above definitions: Points 1, 4, and 5 deal with the boundary, and Points 2, 3, and 6 deal with components and their production. We must now examine whether Zeleny and Hufford's application of these points to the family can be justified without distorting the concept either of the family or of autopoiesis. For each of the six points (except 2 and 3, which will be considered together) I shall give a brief quotation from Zeleny and Hufford (1992a, pp. 155–156) and then my response.

Key Point 1

> The family boundary is usually well defined. The distinction between family and non-family members is rarely ambiguous or subject to fuzzy interpretation. . . . It is not physical. . . . [It] might be defined as the members included in a set. . . . Using "fuzzy" set theory others outside the nuclear family have the potential . . . to be included.

Is the boundary distinction actually well defined? Zeleny and Hufford begin by claiming it is "rarely ambiguous," but how would one classify the following: the cousin who lives abroad and is never seen? the in-laws? the *au pair* here for three months? the live-in nanny? the girl/boy friend (live-in or not)? the divorced wife? the runaway father? What about other cultures where families can take quite different forms (Zeleny and Hufford explicitly have the Western nuclear family as their model)? Is there any reason to suppose that everyone would agree on what constitutes membership? Possibly recognizing this, Zeleny and Hufford then admit that membership may be fuzzy, but if this is the case then there can be no self-defined autopoietic unity.

Key Point 2

> The family system is defined through its clearly identifiable and role-separable components. There are fathers, mothers, children, wage-earners, homemakers, extended family members, aunts, uncles, cousins, "black-sheep, and so on."

Key Point 3

> Family members display system-derived properties that characterize them as family members. Specialization, role-playing, aspirations, preferences, goals, needs, etc. generate interactions which are different from the market-place, church community, or concentration camp.

Here begins the central problem of the paper, a complete confusion of the biological domain with the social domain. In describing components as "mother," "father," etc., Zeleny and Hufford continually confuse the biological individuals with the social roles that may be attributed to them. The difference between the two should be clear with a role like "wage-earner," which may apply to a number of members, but it is equally true of, say, "father," which could apply to a biological father, an adoptive father, a foster father, a stepfather, or a single-parent woman. Moreover, these roles (and therefore the supposed properties that go along with them) are not objective features of the world but social constructs, developed within a particular (patriarchal) culture at a particular point in time and continually negotiated, defined, and enacted in everyday life.

Key Point 4

> The boundary of the family is defined and maintained by the family members themselves. . . . The boundary is maintained through preferential neighborhood relations and interactions between the components (the family members).

Zeleny and Hufford simply state that the boundary is maintained through "preferential neighborhood relations and interactions between components," but what could this mean? First, there are no boundary components, and second, what does "preferential" mean here? It could mean "more frequent," in which case a neighbor may be more a member than a distant relative, or it could mean "preferred," but again friends may be preferred to relatives.

Key Point 5

> The components within the family (the family boundary) are produced through family interactions. . . . Sons are transformed into fathers, fathers into grandfathers, mothers and fathers produce sons and daughters. . . . To become the "head of the family" is an internal social production. . . . Men and women biologically produce children.

This suggestion seems totally confused. As there are no boundary components (Point 1) there clearly cannot be any production of boundary components. They are actually describing the production of nonboundary components, but this notion of production is itself unclear—see Point 6. Nor is it clear that Zeleny and Hufford's boundary could contribute to autopoiesis. A physical boundary has a spatial dimension forming a barrier between inside and outside. This is not the case for a membership-type boundary; some members are not nearer the outside than others. It may be argued that this can be associated with different degrees of membership, but this has been dealt with under point 1.

Key Point 6

> All components of the family, boundary or otherwise, are produced by both biological and social production, as in Key Point #5.

This confusion carries over into the authors' notion of production, where it is not clear if the processes are biological, social, or both. As examples of production processes we are given sons being "transformed" into fathers, "head of the family" as a social production, and men and women biologically producing children. As numerous authors argue (Maturana, 1988; Varela, 1979; Luhmann, 1986; Mingers, 1989; Teubner, 1987), the actual production of biological organisms is a biological process quite independent of whether such organisms then participate in a social family. Maturana, as we shall see in Section 8.4, suggests that social systems are the medium within which organisms realize their structural coupling, not the domain of production of such organisms. It is clear therefore that any attempt to describe social autopoiesis must locate it entirely within the social domain.

Enough has now been said to demonstrate the incoherence of Zeleny and Hufford's attempt to demonstrate social autopoiesis.

8.3.   Varela's Organizational Closure

It should be pointed out that Maturana and Varela themselves have never claimed that social institutions are autopoietic. Indeed, they have not been able to agree among themselves. In the introduction to *Autopoiesis and Cognition* they say that they were going to produce an appendix with their views on the social and ethical implications of autopoiesis but could not agree (Maturana and Varela, 1980, p. xxiv). In the event, Maturana contributed his own views, which were developed later in Maturana (1980b, 1988b).

Looking first at Varela, he clearly recognizes the problems outlined in Section 8.2 concerning social autopoiesis:

> . . . in order to say that a system is autopoietic, the production of components in some space has to be exhibited; further, the term production has to make sense in some domain of discourse. Frankly, I do not see how the definition of autopoiesis can be directly transposed to a variety of other situations, social systems for example. It seems to me that the kind of relations that define units like a firm . . . or a conversation . . . are better captured by operations other than productions. Such units are autonomous but with an organizational closure that is characterizable in terms of relations such as instructions or linguistic agreement. Varela (1981a, p. 38)

He goes on to develop a less specific version of autopoiesis, which he terms organizational closure (1979a, p. 55), which has the same general sense of a closed network of interdependent processes but without the particular specification of physical processes of component production. Such processes could be of many different types, including the nonphysical or symbolic, such as descriptions, ideas, or general computations of any kind. Two examples of physical systems that are said to be organizationally closed but not autopoietic are the nervous system and the immune system. Organizationally closed systems retain most of the important properties of autopoiesis, in particular, autonomy and structure dependence—that is, the sequence of states they follow is primarily determined by their structure and only triggered by their environment. Varela does not, however, specifically develop any social theory.

Varela's later work on the notion of "embodiment" will be discussed in Chapter 11.

8.4.   Maturana: Society as a Medium for Autopoiesis

8.4.1.   Natural Social Systems

Maturana also does not claim that social systems are autopoietic. His approach is to consider what he calls "natural social systems," examples of which are families, clubs, and political parties. He does not make clear what an "unnatural" social system would be, nor does he give a definition or general characterization. Instead, following his own methodology, he sets out to describe a mechanism that would generate the phenomena we would experience when referring to social systems. He answers this question in typical fashion:

> . . . a collection of interacting living systems that, in the realization of their
> autopoiesis through the actual operation of their properties as autopoietic
> unities, constitute a system that as a network of interactions and relations
> operates with respect to them as a medium in which they realize their
> autopoiesis while integrating it, is indistinguishable from a natural social
> system and is, in fact, one such system.
> Maturana (1981, p. 11)

This convoluted phrasing needs a good deal of unpacking, but in essence he means that social systems are not themselves autopoietic, but constitute the medium in which other autopoietic systems exist and interact in such a way that the interactions become bound up with the continued autopoiesis of the components. In other words, a group of living systems (not necessarily human) take part in an ongoing series of interactions with one another. These coordinations of action contribute to the continued survival of the individual autopoietic systems. This generates networks of particular interactions and relations through the structural coupling of the organisms, and these networks become involved in the continued autopoiesis of the organisms. The resulting system (or unity distinguished by an observer), consisting of the living components, their interactions, and the recurrent relations thus generated, is characterized by a particular organization—the social organization. It is also an example of a consensual domain (see Section 6.3). This rather bare characterization can be fleshed out by looking at its consequences:

A. What, if anything, distinguishes social domains from other domains of interaction? Or, put another way, do all autopoietic systems (including cells) exist in social domains? In one paper, Maturana (1980a) seems to say yes, suggesting that even collections of cells in a multicellular organism are an example of a social system. I would not accept this for the reasons put forward in Section 8.2.2, and Maturana himself, in a later paper (1988b), argues that social systems entail a basic emotional attitude toward others of mutual acceptance ("love," for Maturana).

> . . . an observer claims that social phenomena are taking place when he
> or she sees two or more organisms in recurrent interactions that follow an
> operational course of mutual acceptance. . . . The emotion that makes
> possible recurrent interactions in mutual acceptance is that which we connote in daily life with the word love.
> Maturana (1988b, p. 64)

This, it seems to me, implies the possibility of choice—cells cannot choose not to interact with their neighbors—and restricts social systems to those organisms with a high degree of flexibility with respect to their behavior. (In the following discussion I will assume human social systems.)

In fact, Maturana uses the term *social* in a rather specific sense. In saying that social systems are characterized by mutual acceptance, he distinguishes them from other forms of recurrent interactions in which humans participate, characterized by other basic attitudes. Thus, for example, Organizations are not social systems, because they are work communities based on task fulfillment, and military forces are not because they are hierarchical systems based on obedience. This does follow precedents in social science—for example, Tönnies' (1955) distinction between *Gemeinschaft* and *Gesellschaft* and Habermas' contrast of purposive rational and communicative action. However, it seems to me an unnecessary restriction of the term *social*. Organization theorists would argue that there are many different social institutions embodying many different purposes, but no matter how purposeful or hierarchical they are, people's behavior within them is still social. Human beings are social animals, although at times they can act nonsocially (and "inhumanly").

B. A social system, a set of recurrent interactions and relations, is an emergent domain—it is not reducible simply to its participants (see Fig. 8.1). Particular members may join or leave, but the social organization continues. The relationship between people and the social system is circular. The participants, as structure-determined entities, have properties and behaviors determined by their structure. These properties and behaviors realize the particular social systems to which they belong. But this, in turn, selects particular structural states within the participants, as in all structural coupling. In other words, a social system inevitably selects or reinforces behaviors that confirm it and deselects those which deny it.

C. People are members of many different social systems. They may enact them successively or at the same time. These domains all ulti-

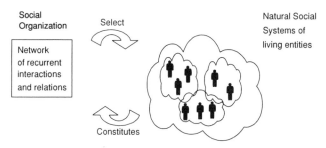

FIGURE 8.1. Maturana's View of a Social System.

mately intersect in the body and nervous system of the individual and may well involve different and possibly contradictory modes of behavior. Membership is very important in human social systems. To become a member means taking on the behaviors appropriate to the domain (consensual coordinations of action), becoming structurally coupled, and then being accepted as such by other members. Decisions about acceptance and rejection are emotional rather than rational and form an implicit boundary for the system.

D. The mutually reinforcing nature described in B) means that a social system is inevitably conservative in the sense that it operates so as to maintain its present organizational relations. Change can come about only through a change in the behavior of the participants—it cannot be imposed by the system. Such change can happen, despite the homeostasis of the social system. An individual may enter a social system and not become structurally coupled to it, instead altering the behaviors of the other members by becoming structurally coupled to them in the course of coordinations of action that do not confirm the social system, or already existing members can reflect upon their experiences in other domains and choose to modify their own behaviors, thus realizing an altered social system. Neither of these, of course, is necessarily very easy. It reminds me of George Bernard Shaw's statement:

> The reasonable man adapts himself to the world: the unreasonable one persists in trying to adapt the world to himself. Therefore all progress depends on the unreasonable man.
> Shaw (1925, p. 238)

E. For humans, interaction is essentially linguistic, and (following Maturana's description of languaging and conversation, outlined in Section 6.3) social systems can therefore be seen as networks of recurring conversations. Conversations are a braiding of language, emotion, and bodyhood, and social interactions involve all three. This is in fact the mechanism whereby the structural coupling of the social system takes place, since linguistic interactions are inevitably physical, involving the body and the nervous system.

8.4.2.   Giddens' Structuration Theory

There are other sociologists whose work does, implicitly or explicitly, relate to autopoiesis, for example, Giddens (1976, 1979, 1984), Cohen, (1989), Touraine (1977), and Bourdieu (1977). We will concentrate on Giddens, who has developed a social theory based on a distinction between "system" and "structure" and what he calls the process of "structuration," by which social structures maintain and produce themselves over time. Although there are significant differences, his work has definite resonances with Maturana's. For example, Giddens' distinction between structure and system is similar to Maturana's between organization and structure although the use of the term structure is reversed. Giddens is concerned with the continual production and reproduction of social structure through time and is strongly antifunctionalist. Giddens himself mentions autopoiesis as being of interest:

> the most relevant sources of connection between biological and social theory . . . concern recursive or *self-reproducing* systems. There are two related types of theory involved here. [The theory of automata] is not of as much interest to the conceptualization of social reproduction as recent conceptions of cellular self-reproduction (autopoiesis) . . .
> Giddens (1979, p. 75)

His work is now well known within sociology, but it is worth examining it in some detail both to make clear the similarities and differences with Maturana and Varela and as a contrast with Luhmann's excessively structural approach (Section 8.5).

The starting point of Giddens' analysis is that society is radically different from nature in that it is a human production, and a central theme in his work is that society is continually produced and reproduced through the skilled performances of actors. In some sense, then, society is seen as a system that continually produces itself, not just through the conscious intentions of members but also through the unintended consequences and unknown determinants of action. In trying to link a theory of action with a theory of structure, Giddens develops three elements: a theory of the subject, an account of the conditions and consequences of action, and an analysis of structure as both constraining and enabling action. I shall look at each in turn, concentrating on structure.

Action is defined as "a stream of actual or contemplated causal interventions of corporeal beings in the ongoing process of events-in-the-world" (1979, p. 55), thereby emphasizing the practical intent of action as praxis, as do Maturana and Heidegger. Such action is always located within three dimensions—time, space, and social structure. He then develops a stratified model of action, or rather of the acting subject, with three levels of consciousness:

Discursive consciousness	Rationalization of action
Practical consciousness	Reflexive monitoring of action
Unconsciousness	Motivation of action

Practical consciousness refers to the normal, everyday awareness that we have of our conduct through reflexive monitoring. It points to the intentional and purposive character of behavior and draws on social knowledge that both enables and justifies the action. Much of this knowledge, however, is tacit—the actor may not be able to formulate and explain it consciously. To the extent that actions can be consciously explained and justified, Giddens talks of *discursive consciousness* and the rationalization of action, while at the other end of this spectrum we have the motivational needs and wants of the actor which may largely reflect *unconsciousness*.

Action, however, goes beyond the conscious and unconscious intentions of the actor to link up with social structure. Actions have unknown and unanticipated consequences, some of which serve to reproduce the social structure. At the same time, it is the social structure that conditions and enables, in an unacknowledged way, future actions. This situation is similar (but not identical) to using a language. In speaking we draw on the already existing structure of a language and, at the same time and by the same act, help to reproduce it.

This leads on to social structure itself, and here Giddens distinguishes among structure, system, and the process of structuration. Taking social structure first, this does not describe empirically observable patterns or regularities as in structural-functionalism but underlying sets of rules that generate the observed regularities, more like classic structuralism:

> Structures exist paradigmatically, as an absent set of differences, temporally present only in their instantiations, in the constituting moments of social systems. [They involve] a 'virtual order' of differences. [They] do not exist in time–space.
> Giddens (1979, pp. 64–65)

This almost matches Maturana and Varela for opaqueness! The term *paradigmatic* stems from the linguistic distinction (originally by Jakobson; see Jakobson and Halle, 1956) between two dimensions of meaning—the *syntagmatic* (metonymical) and the *paradigmatic* (metaphoric) (see Wilden, 1977, Chapter 2). In the syntagmatic dimension, a term gains meaning through its combination with other terms in time or space. In the paradigmatic dimension, it gains meaning by its selection from other possibilities within a code or set of rules. For example, in the sentence "fetch me the hammer," the word "hammer" is combined (syntagmatic) with "fetch me the" (it could have been "where is the hammer" or "hammer in the nail"), and selected from other possibilities (paradigmatic) such as "screwdriver" or "cup of tea."

Structure is thus seen as similar to a code or set of rules that governs possible selections of social action. It is constituted as an "absent set of differences," again referring to the linguistic idea that the elements of a language can be characterized only by their difference from other elements, not by reference to external reality (Saussure, 1960). The rest of the definition emphasizes that structure is not empirically observable as such but is exhibited only in particular social interactions. In fact, structure should really be seen as the structuring properties of social systems—"these properties can be understood as rules and resources, recursively implicated in the reproduction of social systems" (Giddens, 1979, p. 64).

Moving now to social system, this is more straightforward. This refers to the empirically observable regularities and relationships among actors and groups through their social actions and practices. "Social systems, by contrast to structure, exist in time–space, and are constituted by social practices. The concept of social system . . . refers to reproduced interdependence of action" (Giddens, 1979, p. 73). Social systems therefore display a syntagmatic patterning in time and space. This observable patterning comes about through the virtual (unobservable) structure governing their activity. Giddens also refers to institutions as parts of social systems. By this he means particular practices that are widespread in both space and time ("deeply sedimented," to use his phrase). This would include traditional social entities such as the family, but also more diffuse customs and practices concerned with, for example, sports, working, politics, and social customs.

The relationship between system and structure is provided by the concept of structuration, a twofold process, which Giddens sometimes refers to as the duality of structure (Fig. 8.2). First, structure organizes the practices that constitute a social system—actors draw on the struc-

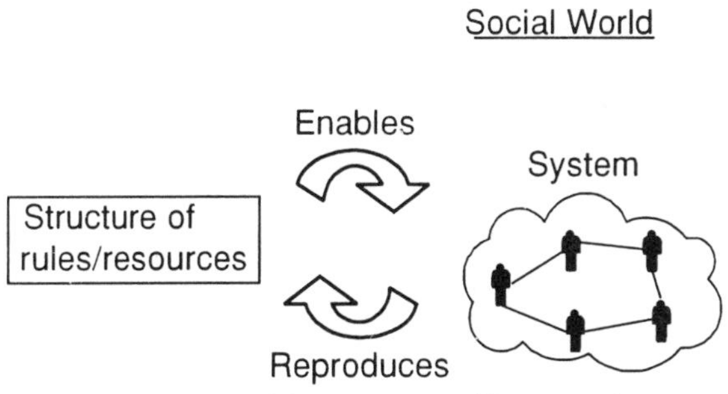

FIGURE 8.2. Giddens' Process of Structuration.

tural rules and resources in the production of interaction. But second, it is precisely and only these interactions that reconstitute (and possibly modify) the structure. "The structural properties of social systems are both the medium and the outcome of the practices that constitute those systems" (Giddens, 1979, p. 69).

In examining structuration, we can focus our attention either on the way actors draw on the social structure (Giddens terms this *strategic conduct*) or at the way structural rules are reproduced through social interaction *(institutional analysis)*. Cohen points out (1989, p. 89) that it is also necessary to analyze the pattern and development of social systems *(systems analysis)* in themselves, bracketing both their reproduction of structure and actors' employment of structure. I might add that one could also analyze (although it might actually be the same thing) the patterning and development of structure, separate from systems.

Finally (in this brief introduction to Giddens' work) he distinguishes three dimensions of social practices, which are reflected at both the structural and interactional levels. These three are *signification* (and *communication*), *domination* (and *power*), and *legitimation* (and *sanction*). These are not separable but are three different aspects of practice. Signification concerns the production of meaning in interactions. Structurally it involves sets of semantic rules, and interactionally it relies on intersubjectively shared interpretive schemes. Legitimation concerns not what acts mean, but whether or not they are appropriate or acceptable. It involves moral and social rules structurally and norms and sanctions interactively. Domination refers to the structuring effects on interaction of inequalities of resources. At the interactional level it emerges as the power of agents to bring about desired consequences.

8.4.3. *Giddens and Maturana*

Structuration and autopoiesis certainly have a number of *prima facie* resonances.

1. Both place the production and reproduction of systems at the center of their theories, in particular the idea that systems can be recursively self-producing.
2. Both maintain that explanations should be nonfunctionalist and nonteleological but concern the actual historical interactions and processes that have occurred.
3. Both draw a clear distinction between that which is observable, having space–time existence (structure for Maturana, system for Giddens), and that which is not but is still implicated in the con-

stitution of the system (organization for Maturana, structure for Giddens). Giddens' term *virtual* seems a good one here.

4. Both take an essentially relational view and distinguish the same three types of relations (see Section 2.3.2): constitution/space, order/time, and specification/paradigm.

5. Both stress the primary nature of action as involved, practical activity in the world and do not try to divorce thought and language from bodyhood (or emotion, in Maturana's case).

6. Both recognize the intersubjectivity, the shared, preexisting meanings (the consensual domain) involved in social interaction.

However, on closer examination a straightforward assimilation of autopoiesis to structuration theory is not possible, for reasons to be outlined below. The comparison should be, not with autopoietic structure/organization but with Maturana's social theory.

Can Giddens' "system" be taken as an example of Maturana's and Varela's "structure"? It is observable, existing in time–space, constituted by particular social practices, institutions, and their relations and reproduced through time. The first question is, what are its boundaries? How is it distinguished as a unity? Does it correspond to a society, a nation, a culture, a particular economic system (e.g., capitalism), or what? As a unity, what would its properties and its domain of interactions be? Maturana and Varela argue that a unity is distinct from its components—it has emergent properties. Second, can social practices be seen as processes of production? Practices certainly interact with one another across space and time, but it is not clear that they actually produce one another. This would need, at the least, a good deal of exemplification and clarification.

Moving to the comparison between Giddens' "structure" and Maturana's and Varela's "organization," again the surface similarities disappear. As Maturana and Varela use the term "organization," it is simply descriptive of a particular abstract concept, a particular combination of relations. It has no causal or enabling powers; it does not generate the particular behavior or properties of the unity's components. In the case of physical autopoiesis, the components have properties and behaviors that follow physical laws. Giddens' structure is quite different. It is not so much a descriptive concept as a causal form that may not have physical existence but nevertheless permits activity within the system.

It seems to play a role similar to that of natural laws in the physical system (see Fig. 8.3), determining the properties and behaviors of the social system. "Determining" is not quite the right term—it should be

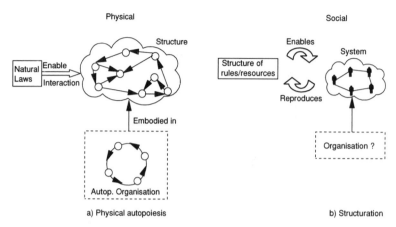

FIGURE 8.3. Physical Autopoiesis vs. Giddens' Social Structure.

"allowing" or "enabling." Social structure does not determine what *will* happen but what can happen.

Chess provides a reasonable analogy to illustrate the difference between the terms. The game relies on a set of rules that both allow and constrain what can happen. These rules do not essentially have physical existence, although they are written down. Besides the formal rules, there exists a large body of generally accepted knowledge, known to a greater or lesser extent by various players. The rules and knowledge are equivalent to Giddens' structure; they enable the game to occur. In contrast, an illustration of Maturana's organization might be a particular combination of relations between pieces—for example, stalemate by repetition or perpetual check. The organization itself has no force, but if such a combination occurs then, because of the rules, certain consequences follow.

Thus, I would argue that Giddens' *structure* is not similar to Maturana's *organization*, but this does leave open the possibility, as shown in Fig. 8.3, of demarcating an organization appropriate to social systems. Luhmann's autopoietic social theory can be interpreted in this way (see Section 8.5).

As mentioned above, a more fruitful intersection is with Maturana's analysis of social systems. As we have seen, for Maturana social systems are not autopoietic but are the medium for interaction and structural coupling constituting a social organization. They are consensual domains consisting of networks of recurrent interactions and relations. They are separable from the people who realize them in that people may come and go but the social organization carries on. This seems to me highly compatible with Giddens' "system" and "structure" (see Figs. 8.2 and 8.3). Maturana's natural social systems are Giddens' institutions within the social system, and Maturana's *social organization* is Giddens' *structure*. Both envisage similar closed relations between the two—for Giddens, system interaction reproduces social structure, which enables interaction; for Maturana, system interaction constitutes social organization which selects interaction.

I would suggest that Giddens' analysis provides a more detailed picture of the consensual domain of social organization. It specifies that the constituents are not just interactions and relations in general but, specifically, social practices in the system and rules and resources in the structure. Moreover, he suggests the main dimensions of these interactions: signification, power, and legitimation. On the other hand, I think Maturana's concept of structural coupling and his explanation of the biological foundations of language and social interaction through the nervous system and bodyhood provide useful support for structuration theory.

8.5.   *Luhmann's Autopoietic Society Based on Communications*

Having discussed Maturana and Varela, who do not say that society is autopoietic, we now move to someone who claims that society is the production not of physical entities but of communications. Niklas Luhmann is one of the major German social theorists of the last thirty years. He has had a large and wide-ranging theoretical output developing a detailed theory of social systems in general and modern society in particular. This is founded in general systems theory and particularly the work of Parsons, who tried to reconcile action theory with systems theory. Luhmann's use of autopoiesis (from the early 1980's) does not represent a radical departure for him; rather, he has been able to incorporate the ideas quite neatly into his own theory of the differentiated society.

His work up to autopoiesis is well covered in *The Differentiation of Society* (1982a), while his major work developing autopoiesis is *Soziale Systeme* (1984a). There are a number of papers outlining, in a fairly schematic way, his use of autopoiesis (Luhmann, 1982b, 1983, 1984b, 1985b, 1986, 1987a, 1993) and, most usefully, a book—*Ecological Communication* (1989b)—in which he uses his theory to analyze the ways in which ecological problems may or may not impinge on society. He has also produced an analysis of the nature of risk in society (1993a,b) based on autopoiesis. He has participated in major debates with Habermas, documented in Luhmann and Habermas (1971) and most of Habermas' other works, especially Habermas (1990). Bednarz (1988) and Geyer and van der Zouwen (1991) argue for Luhmann's development of social autopoiesis, Blom (1990) provides an informed review of Luhmann's concept of social structure, Gastelaars (1992) applies autopoiesis to the sphere of public prevention (government activities intended to prevent specific diseases), and Kickert (1993) to public administration. Luhmann's particular application of autopoiesis to the law has itself generated a major debate, which will be covered in Chapter 9.

Grasping Luhmann's ideas poses problems similar to those involved in interpreting Maturana's writing. He uses common words (e.g., "communication") in special ways which are only briefly explained and his writing is highly abstract, offering very few illustrations or examples.

8.5.1.   The Differentiation of Society

It is best to begin with the fundamental driving force of his theory, the idea that modern society is necessarily becoming a functionally differentiated one. (The particular nature of "society" for Luhmann will be brought out later.) Society, as a system, is faced by an environment (all that is not society, not just the physical environment) that is inevitably more complex than itself. For it to survive, it must somehow match its variety to that of the environment. (This is essentially Ashby's (1965) "law of requisite variety.") There are only two ways to go—it must reduce the complexity of its environment, or at least that which it experiences, by isolating itself and reducing its possible interactions, or, it must improve its own variety and become increasingly complex itself. Society, like living systems in general, has followed the latter route and become increasingly differentiated internally.

Internal differentiation means simply that the system develops a greater number of its own subsystems. This can occur in different ways and has done so through the evolution of society (Luhmann, 1982a, pp. 232ff). Initially, it was on the basis of *segmentation*—the generation of many essentially identical subsystems, such as villages in the middle ages. The next stage was differentiation by *stratification*, by which society split into unequal subsystems forming a hierarchy such as peasants, clergy, and nobility. Modern society can be seen as a development from stratified to *functional* differentiation: subsystems become established by the particular tasks that they carry out—for example, the economy, politics, law, science, education, and religion. These subsystems become highly autonomous, distinguishing themselves from their environments self-referentially. Society no longer has a center or controlling subsystem, but becomes the indeterminate outcome of the interactions among these independent but interdependent domains.

Luhmann brings in autopoiesis by arguing that all these subsystems and society itself are autopoietic unities and are thus organizationally closed and self-referring. In doing this, he recognizes the problems in defining social autopoiesis, in particular the exact nature of the components and the processes of their production (Luhmann, 1986, p. 172). He is happy to accept that social systems do not consist of or produce the (physical) people who participate in them (indeed, he argued this before he made any mention of autopoiesis; see Luhmann, 1982a, p. xx). So in what sense can they be autopoietic? His answer is similar to Varela's in suggesting that there can be closed, self-referential systems that do not have physical production as their mode of operation. These include both social systems and psychic systems (human consciousness). He differs from Varela in that he calls all such systems autopoietic, whereas Varela restricts that term to living physical systems.

8.5.2. *Autopoiesis as the Production of Communications*

So what are the elements of social systems that continually produce themselves? Not conscious thoughts, nor behavior or actions, nor even language, but communications, or rather *communicative events*.

> Social systems use communications as their particular mode of autopoietic reproduction. Their elements are communications which are recursively produced and reproduced by a network of communications and which cannot exist outside such a network.
> Luhmann (1986, p. 174)

Each subsystem defines for itself what is and is not a communication for it and then consists of networks of particular communications, which always refer to previous communications and lead on to other ones. Society as a whole encompasses all the communications of its subsystems.

It is important to understand what Luhmann means by communications since he uses the term in a very specific sense. He stresses that it is not what we might normally mean by a communicative act, such as a statement or utterance by a particular person. Indeed, it is at a different level from people and their thoughts and actions. For Luhmann, these are not part of the social system at all, but its environment. He characterizes a communication as an event consisting of three indissoluble elements—information, utterance (communication or action), and

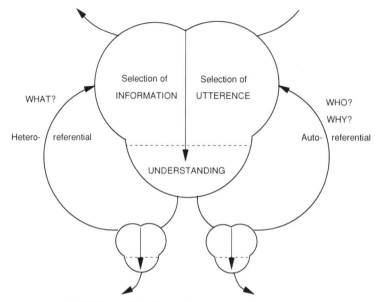

FIGURE 8.4. The Three Elements of a Communication.

understanding (comprehension)—which can enable further autopoietic operations to occur. The terms in parentheses have also been used by Luhmann. Each of these elements is said to be a selection, that is, one possibility chosen (but not necessarily by a person) from many. It is the operation of the autopoietic system which defines and makes the selections (see Fig. 8.4).

Broadly speaking, information is what the message is about; it is the difference that makes a difference (Bateson, 1979). Utterance is the form in which it is produced—how? By whom? When? And understanding is the sense or meaning that it generates (which can include misunderstandings) in the receiver.*

---

*Luhmann developed these categories from a typology by Buhler (see Luhmann, 1986, p. 188, note 2). Buhler's scheme (1982, p. 147; see also Habermas, 1992, p. 57), published originally in 1934, was a classification of the three relations that a linguistic expression may enter. It has, first, a relation of expression to its sender or communicator; second, a relation of representation to objects or states of affairs; and third, a relation of appeal to its receiver. These, then, are supposed to correspond to Luhmann's utterance, information, and understanding, although it seems to me that utterance includes both the relation and the actual linguistic expression. Note that Luhmann is critical of the traditional transmission model of communications with self-conscious senders and receivers (as is Maturana—see Chapter 6).

There must be at least two parties involved, the communicator and the communicatee (my terminology), but these may well not be individual people. All these elements are generated or coproduced together as a unity, and this event allows for the possibility of further communications. It is important to stress that all three aspects are distinctions made by the system—the system determines what is information for it, how it may be embodied, and how it may be interpreted. This is the closure of autopoietic systems.

Of the three, understanding stands in a special relation to the other two. Understanding draws the distinction between information and utterance (Luhmann, 1982, p. 183) and recognizes that they are selections in different dimensions. Information is the "what" of the communication—it is produced by the system out of the perturbations the system undergoes, and the system determines whether it originates or refers outside the system to the environment. The utterance is the "why now," the "how," and the "who" of the communication and so is inevitably self-referential. Again, these distinctions are made by the communication itself, which is attributed to an agent rather than being the conscious production of an agent. It is this distinction between information and utterance that allows for an arbitrariness between the two and provides the possibility of further autopoietic production, for without it understanding would simply be perception rather than communication. This distinction between information and utterance is very similar to that between the propositional and illocutionary content of utterances analyzed by Austin (1962), Searle (1969), and Habermas (1984).

Luhmann gives almost no actual examples of communications, but I would suggest the following within various subsystems. In the law, a legal communication might be the judgment of a court. It contains a particular selection of information (the nature of the case, the main considerations, reference to laws and previous decisions—earlier communications); it is presented in a particular way (a speech, a written judgment); and it is interpreted in particular ways. The judgment as a whole leads to further communications, both directly through its consequences and indirectly as part of case law. In science, a communication is about a new theory or the results of a particular experiment. Again, it is a particular selection of information (reference to previous results, definitions of what is or is not acceptable); it is presented either in a paper in a journal, or at a conference, or at a demonstration; and it is understood in certain ways and leads to further communications. In the everyday world of work, a communication may be a memo or phone call and the understanding it generates in the recipient. It may provoke a further communication either in reply or to others. Even the absence of a reply may be taken as a communication and may generate further ones.

We can visualize the whole system as an ongoing network of interacting and self-referring communications of different types and see how they can be separated from the particular people involved. The people will come and go, and their individual subjective motivations will disappear, but the communicative dynamic will remain.

Communications, Luhmann argues (1986, p. 177), are not the same as individual communicative acts, but are more fundamental. This is because, first, actions need not be inherently social whereas communications are social, although this does verge on the tautological since for Luhmann the social is defined as a system of communications. Second, social actions already presuppose communications in the sense that they rely on or raise the expectation of recognition, understanding, and acceptance by others. In other words, a social action is inevitably already a communication. Yet, third, a communication is more than simply an action. It involves and therefore includes the understanding of another party and so goes beyond the individual action to form the link necessary for social operations. A communicative act in itself leads to nothing; it is only when it generates some understanding in another that it can trigger a further communication.

Having looked at the structure of communication, how are the dynamics of autopoiesis constituted? Essentially it is a network through time of communications referring to other and past communications and leading to new ones (Fig. 8.5). However, it is quite a different form of production from physical autopoiesis, for communications are events. They occur at a point in time and then disappear. They may leave

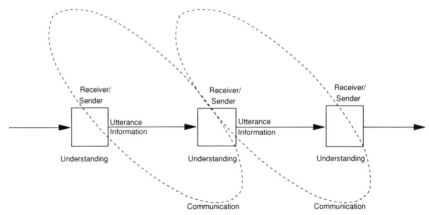

FIGURE 8.5. Communication as an Ongoing Process: Synthesis of Utterance, Information, and Understanding.

traces—memories, papers, films—but these are not the events themselves. What is vital is the generation of the next (communicative) event, for without this autopoiesis stops, and that this event is different from the previous one. So communicative autopoiesis is not a production of structure or pattern or repetition but of networks of differentiated events.

Looking specifically at the relationship between communications, Luhmann suggests (1986, p. 175) that they will be either heteroreferential or autoreferential (Fig. 8.4). A later communication distinguishes between the information and the utterance of an earlier communication. It then concerns either (hetero-) the information (questioning it, denying it, or enlarging on it) or (auto-) the utterance (asking how it was said, why it was said, or who said it). In each case, the later communication makes its own particular distinctions (or selections) among these varied possibilities.

We can see here the relationship of communication to meaning (Luhmann, 1985b, p. 7; 1990a). Events (especially communications) refer to and are related to many other events and possibilities. The production of communications is precisely this selection from the manifold possibilities—distinguishing what it is by what it is not. It is these related events and possibilities that constitute meaning. Meaning is the openness of all possibilities—all the relations, distinctions, and denials that could be generated—a very Husserlian construction. It is that which provides the newness of and difference between communications. On the other hand, a particular communication closes this off; it fixes one possibility in order that something might actually happen. Autopoietic communication can thus be seen as *meaning-processing* (Luhmann, 1989b, p. 17), generating distinctions to convert the open field of meaning into the particular information or utterances that thereby constitute a society. For a related but not identical analysis of information and meaning based on autopoiesis, see Mingers (1993a, 1993b) and Kampis and Csanyi (1991).

8.5.3.   *The Autopoiesis of Society*

> A social system comes into being whenever an autopoietic connection of communications occurs and distinguishes itself against an environment by restricting the appropriate communications. Accordingly, social systems are not comprised of persons and actions but of communications.
> Luhmann (1989b, p. 145)

Having grasped Luhmann's ideas about how social autopoiesis actually operates, we can now return to see society and its subsystems work as a whole (Fig. 8.6). As explained in Section 8.5.1, society differ-

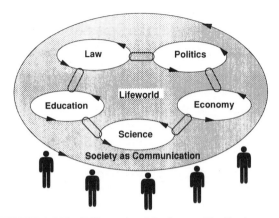

FIGURE 8.6. The Differentiated Society and Its Environment.

entiates itself into subsystems each of which is an autopoietic network of recursive communications. Society itself is also autopoietic, consisting of all these communications plus all others not specifically involved in subsystems—the communications of the "lifeworld"—the everyday world of action and communication (Luhmann, 1989b, p. 36). As such, it distinguishes itself from its environment—that which is not communication. Thus, not only the physical environment but also people and their consciousnesses are in the social system's environment. Society is a closed system in that it cannot communicate directly with its environment since the environment, by definition, does not communicate. Events happen in the physical world (e.g., pollution), but this does not affect society until it becomes the subject of a communication—"society cannot communicate *with* but only *about* its environment according to its capacities for information processing" (Luhmann, 1989b, p. 117). This does not mean that society is totally isolated—it is like examples of physical autopoiesis, organizationally closed but interactively open. The environment (especially people) can trigger or irritate society, and society may then generate a communication, but its nature and form will be determined by society or a particular subsystem, not by the environmental disturbance.

In a similar way, the subsystems also distinguish themselves within society and specify their own internal environments. They too form closed networks of communications, each one being able to process or deal with communications of its own type only. Luhmann analyzes their workings with the terms *codes* and *programs.* He argues (1989b, pp. 36ff) that each subsystem utilizes a particular binary code representing the good/bad or positive/negative for that subsystem. For example, the code for the law is legal/illegal, for the economy to pay/not to pay, for science truth/falsity, for politics the holding/not holding of office. The code provides the basic guidance for a subsystem, for without it the self-referential operations would be entirely undetermined. The code itself is just the particular categories, and it therefore requires some means or criteria for assigning events to a category. This is the program—the rules for coding. This separation is one way in which variety can be increased since it allows the program to be changeable even though the code is not. For example, the historical development from natural to positive law involves a shift from defining legality by religious or natural criteria, which were unchangeable, to criteria defined by society, which are thus open to change.

For subsystems, the other functional subsystems exist as part of their environment and there are much greater interactions and dependencies among subsystems than between society and its environment. The subsystems have become autonomous and independent, but at the same time more interdependent since they rely on the existence of the other subsystems to carry out particular functions. Interactions among subsystems are reasonably well defined; legal communications can give rise to economic ones, which in turn can trigger political ones. When a subsystem is triggered by its environment and generates a communication about a particular matter, this is called *resonance*. Going further, Luhmann uses the metaphor to describe the resonance of society as a whole, that is, the subsystems with each other (Fig. 8.6).

For example, a discovery in science such as genetic manipulation may lead to political decisions about limits of acceptability, resulting in new laws governing its use and, eventually, economic activity; each of these feeds back to the scientific subsystem. Moreover, these effects may be disproportionate among subsystems; a minor occurrence in one may trigger a major response in another.

In a recent paper (1992), Luhmann brings in the idea of structural coupling, a concept conspicuous by its absence previously. He interprets this (correctly, I think) as the way in which the system's structure develops to presuppose or expect that certain perturbations will occur in its environment. In the social case, this will be embodied in the particular expectations of communication—expectations that, for example, there will be consciousness and activity. These perturbations (or irritations, as he calls them), do not, of course, exist as such in the environment. The system determines what may and may not be a perturbation.

In any event, he suggests that there are several domains of structural coupling. There is, first, the coupling of communication (i.e., society) to sense-systems, that is, individual consciousness, but not, he argues, to the general physical world. Then there is the coupling of subsystems to society itself. Indeed, this coupling is the closest, since the subsystems are not something other than society but part of its very constitution. Nevertheless, they do distinguish themselves through their operation. Finally, there is structural coupling between subsystems, and here he details a few specific combinations. The economic and legal subsystems are linked mainly through the law of property and contract, and politics and the legal system by constitutional law. Events in these fields generate communications in both the connected subsystems, which then become part of the subsystems' autopoiesis. Although he does not suggest it, it seems to me that such structural coupling can go some way toward generating stability among the otherwise autonomous and uncoordinated subsystems.

We are thus left with a view of society very different from the traditional one. Society is essentially centerless—there is no core or fundamental division driving it, and there is no privileged position from which a rational overall view can be developed. Instead we have self-defined autonomous subsystems in a constant process of renewal and redefinition, locked together in a fragile balancing act, resonating among themselves but relatively unresponsive to society's external environment.

8.5.4. Conclusions about Luhmann

Luhmann's work represents a very thorough and detailed, not to say radical, social theory, but this is not the place to debate its important implications for theoretical sociology (see Habermas, 1990, pp. 368–386) or policy science (where there is a detailed analysis of autopoietic theory and societal steering; see Veld *et al.*, 1991). Rather, I will restrict myself to points concerning the use of autopoiesis and a few more general considerations.

For autopoiesis this is a bold attempt to define an autopoietic unity in the nonphysical domain. It defines the basic components of such a system—in this case, communications—and holds consistently to this without confusing domains by, for example, including people within the system. Production is shifted to events rather than material components. Finally, the circular and self-defining nature of the production network is brought out well, as is the combination of organizational closure and interactiveness.

Other aspects are less successful. First, the problem of boundaries is not properly resolved. For Luhmann, systems differentiate themselves by defining their own communications, so there is at least a distinction between that which pertains to the system and that which does not, but that is not the same as a boundary consisting of particular boundary components. For this reason, I would agree with Varela and say this is a case of organizational closure rather than autopoiesis, reserving the latter term specifically for systems that meet the full definition. Maturana (Krull *et al.*, 1989) also does not wish to characterize social systems as autopoietic. He recognizes the possibility of there being an autopoietic communication system, consisting of a network of production of communications, but argues that the organization of a social system is different, consisting of a network of human coordinations of action (see Section 8.4).

Second, Luhmann does not appear to use the distinction between organization and structure, yet this distinction seems potentially very useful. If we take the law, there are many different legal systems in the world, and any particular one changes over time. We thus observe many different structures but they all, presumably, embody the same basic organization of closed communication. Luhmann (Krull *et al.*, 1989) says that he did not use the term *organization* because it already had a particular sociological meaning.

Third, there is a question as to the possibility of autopoietic systems emerging or developing within an already existing autopoietic system. Biologically, we can see that second-order autopoietic systems may develop from the coupling of first-order ones (see Section 3.3.1). In Luhmann's terms, this would be equivalent to the various subsystems establishing themselves first and then coming together to compose society. This is clearly not what happened, and the problem is, is it actually possible for it to happen that way? Can there be self-differentiation into autopoietic subsystems?

Fourth, there is no significant attempt to show how societal communication, as an independent phenomenal domain, emerges from the interactions of the human beings who ultimately underpin it. Without human activity there would be no communication. Maturana is always careful to show how new domains arise out of the interactions of observers, but with Luhmann the observer is lost completely in favor of the observation. This is an important lacuna with a number of concomitant problems, some of which are also identified by Habermas. How do communications actually occur? It is one thing to say analytically that communications generate communications, but operationally they require people to undertake specific actions and make specific choices. Is not the claim that "communication produces communication" stretching the notion of production too far? One communication may stimulate another, but surely it does not *produce* or *generate* it (see also Teubner, 1993, Chapter 2). How does this interaction occur? What factors affect the selections that are actually made? In general, what is the relationship between the psychic systems of individual consciousness and the social system of communication?

Luhmann's theory would seem to rely on his concept of meaning as the link between the two. A communication opens up possibilities through its meaning to people whose selections then generate new communications. However, this appears to be a very individualistic analysis. No attention is paid to the constituting of these subjects as subjects and the major role of language. In Maturana's terms, Luhmann ignores the importance of languaging and consensuality, which provide an already existing *a priori* structure generating an intersubjective domain of interactions. Such an approach would provide a general bedrock in the lifeworld out of which the particular specialized communicative subsystems can be seen to arise.

Fifth, there is a problem with Luhmann's avowed functionalism. Luhmann specifically uses the term "functionally differentiated society" and appears to argue that this has come about in response to the complexity of society's environment. This certainly appears to be an adoption of functionalism, which Maturana and Varela (and Giddens) strongly reject. However, it seems to me that it might be possible to reformulate Luhmann's work more in terms of structural coupling and thus show nonteleologically how society's particular structure has come about.

Finally, I find it difficult to accept the restriction of the social purely to communication and the supposed separation and purity of the various subsystems. In the economic sphere, for example, it may be that the ultimate operations underlying it are monetary payments, but if we seek to explain particular happenings we immediately find that social, political, and legal factors are at work. This is because it is people who make economic decisions—to buy or sell or invest or lend—and people form a nexus among all the different subsystems. Their decisions are affected by their expectations, which are conditioned by communications from other domains. Moreover, communications can often be said to belong to more than one domain. For example, signing a loan agreement both generates a transfer of money and establishes legal obligations. A research grant is both a communication about academic status and a payment.

8.6.   Autopoiesis as Metaphor

Finally, we will look at the work of Morgan (1986, Chapter 8), who does not actually claim that Organizations are autopoietic, but that autopoiesis is a useful metaphor in thinking about how Organizations work. He develops three ideas.

First, the relations an Organization has with its environment are not determined so much by the environment as by the Organization and its own internal self-image or identity. Organizations are continually concerned to re-create and maintain their image and identity by projecting themselves onto their environments, and what they monitor in the environment is, in turn, a reflection of their own concerns and interests. This is more than merely choosing to observe or not observe certain things but reflects the idea that it is the structure of the system that determines the interactions the system can or can not have. Organizations have their own languages and literally cannot interact with something unless it fits into their perceptual categories.

Second, Morgan identifies "egocentric" Organizations, which are overly concerned with maintaining their current identity despite it being inappropriate to their environment. Examples are companies that try to stick to their traditional practices despite changes in technology (e.g., watchmakers) or companies whose activities alter their environments to their own longterm detriment (e.g., through pollution). Rather, they should be aware that their structure must be one that allows structural coupling to their environment and that structure can, if necessary, be changed without a loss of identity.

Third, in looking at the history of change and development of an Organization, it should be concerned that its identity is flexible and compatible with its environment and context:

> The theory of autopoiesis suggests that the pattern of organization that evolves over time does so in an open-ended and evolving way. Some forms disappear and others survive through transformations controlled by the self-referential processes that define the total system.
> Morgan (1986, p. 247)

Because Organizations are structurally determined and can only be triggered by the environment, their structure may fail to develop in ways necessary to maintain their autopoiesis, and they may discover this only when it is too late.

In overall terms, using autopoiesis metaphorically is reasonably unproblematic—one does not have to agonize over the deep ontological problems. Equally, however, the results are merely metaphoric and have no greater claim on our attention.

8.7.   Conclusions about Social Autopoiesis

In this chapter I have explained both the attractions of characterizing social systems as autopoietic and some of the problems, and I have explored a range of responses to these problems. I believe that, at the moment, social autopoiesis remains a highly debatable possibility, no more. I agree with Varela that applying autopoiesis in a strict sense, to include notions of production and boundaries, to social systems cannot be sustained, but that some more general idea of closure may well be applicable. The problem is to determine unambiguously the nature of the components, processes, and relations of such a system.

Luhmann's work, taking the components to be communications, is the most highly developed autopoietic social theory but is based on a number of questionable premises, particularly the relations between communication and actors. On the other hand, Giddens' structuration theory addresses the action/structure relationship and seems quite compatible with Maturana's work, but needs reworking within an autopoietic or at least closure framework. Between these two, I feel that there is space for interesting developments to take place.

However, the question is particularly difficult because the nature of the social domain itself is extremely controversial. There are philosophical disputes over the epistemology of natural science, but at least this is concerned with a domain of physical existence. How much harder it is to grapple with the social domain, where there is dispute over the nature of social "objects" and even their very existence, and where we are ourselves observers from within.

9

Law as an Autopoietic System

9.1.   Introduction

The application of autopoiesis to law has developed from Luhmann's work on social autopoiesis, described in Chapter 8. As we saw, he analyzed society into a number of different functional subsystems, each based on a particular mode of communication. One of these was law. Luhmann had developed a sociology of law, published in 1972 as *Rechtssoziologie,* before he utilized autopoiesis. An English translation appeared in 1985 *(A Sociological Theory of Law),* and this contained an extra chapter incorporating the basic notions of autopoiesis. Since then Luhmann has written further papers (1985c, 1987b, 1987c, 1989, 1992a,b); the idea has been taken up especially by Teubner (1984, 1987b, 1987c, 1989, 1990, 1993) and has generated a good deal of controversy (Teubner 1987a; Cardozo Law Review 13(5), 1992; Amselek and MacCormick, 1991) within law. Teubner has applied it to the problems of legal regulation (1985, 1987d), Power (1994) to accounting and the environment, and King and Piper (1990) have explored this characterization of law in a book about how the law "thinks" about children.

9.1.1.   The Attractions of Autopoietic Law

To begin with, what is the attraction of thinking of applying autopoiesis to law? In broad terms, it is similar to the reasoning for Organizations and social systems in general—autopoiesis might explain some of the apparent characteristics of legal systems. In particular, the developing autonomy and even self-reference of legal systems combined with their increasingly dynamic and changing nature.

Historically, there has been an evolution within legal systems from "natural" to "positive" law (Luhmann, 1982a, p. xxxiii and p. 158ff). Originally, the law merely reflected or enacted norms and values which were seen as natural—independent of people and societies. These norms and values stemmed from God, through religion, or just seemed to reflect the natural order of the world. The point was that, law did not determine right and wrong, merely enshrine it. However, with the loss of belief in a natural order or religious dogma, the law has been increasingly seen as an internal mechanism for mediating conflict within society. Decisions about right and wrong, legal and illegal, are societal decisions, and law itself becomes involved in positively specifying this. This allows for a greater contingency; the legal system develops formal methods for making and altering law rather than relying on externalities; it permits a separation of law and morality (laws could be enacted which are judged to be immoral); and legality is seen to reside simply in a law being legal (i.e., having been created following the correct legal procedures). Note, however, that there is still debate among legal theorists as to whether law does have some minimum rational or moral content (Finnis, 1980).

This development of positive law has led to it being seen as becoming both more autonomous and more dynamic and changeable. Autopoiesis provides a mechanism that generates both these characteristics (Jacobson, 1989; Van Zandt, 1992). The law in most modern states is in a constant state of change with new and ever more complex statutes being produced and new and changing interpretations being created through case law. This occurs partly because of the severing of law from unchanging natural presuppositions mentioned above, and partly because of the increasing pace of change in society as a whole, and particularly its scientific and technical domains.

> The legal systems of advanced industrial democracies constantly generate and transform law in every legal act and communication. A perpetual motion of norms sharply distinguishes certain legal systems.
> Jacobson (1989, p. 1648)

Thus a characterization of law must be found that can encompass both the ongoing unity and coherence of a legal system and its constant change.

The second effect of the positivity of law is increasing autonomy. Legal matters, rather than being spread throughout society, all coalesce within the one system. The law becomes responsible for both defining and implementing legality and illegality. This is done increasingly with reference to itself—that is, its own regulations and norms for changing the law and its own previous decisions. Laws have force only if they have been enacted legally; law has become self-legitimating (Lempert, 1987). This can be seen with reference to Hart's (1961) distinction between primary and secondary rules. The primary rules of a legal system are accepted as binding because they have been generated and applied in accordance with the secondary rules. These secondary rules are, of course, also created through the legal system.

This has generated a longstanding debate within legal theory (Ewald, 1987) as to whether ultimately law is self-dependent (the pure theory view) or is determined externally by society (the sociology of law view). To some extent, autopoiesis could be seen as a possible way of reconciling or overcoming this split by envisaging the relations between an autonomous law and its encompassing society in a different way. That said, however, autopoiesis does at first sight seem to stem much more from the pure theory camp, particularly the work of Kelsen (1967). He argued that law needs no outside foundations, because the very positivity of law provides its own foundation and the law itself regulates its own creation, a process "in which the law, so to speak, unceasingly creates itself" (Kelsen, 1967, p. 299 and p. 318, quoted in Ost, 1987). This clearly foreshadows the self-producing nature of autopoietic systems, although Kelsen perceives there to be a hierarchical structure of norms and values rather than a circular one.

Following from the autonomy of law is the fact that its self-reference can generate paradoxical situations that legal theorists find hard to account for (Teubner, 1990, 1993). For example, is the most basic distinction between "legal" and "illegal" itself "legal"? If we say it is legal, we have a simple tautology, but if we say it is illegal, then we have a self-negating paradox like the sentence "this sentence is false." If the distinction is illegal, then we should not make it, so it is not illegal, and so on. There are many other examples. As we have seen, laws are legal only because they conform to other (secondary) laws which must themselves be subject to their own prescriptions. What if laws were passed which referred to one another but were contradictory or in conflict? Hofstadter (1980, p. 692) gives an example: during Watergate, the president threatened to obey only a "definitive" ruling of the Supreme Court, then argued that only he could decide what was "definitive." And while I write this, President Yeltsin in Russia is claiming powers over the Congress, while the Congress denies that such powers are legal. Even a single law could be self-contradictory—for example, a law passed by a simple majority specifying that all laws retrospectively need a two-thirds majority. Then there are situations in which illegal acts bring about changes in the law, which then legitimize the acts—for example, resistance to laws that then brings about their repeal, or revolutions and coups d'état that create new laws.

Finally, Teubner lists a number of concrete examples which might lead to paradoxes:

> 'Who watches the watchman' as the problem of constitutional jurisdiction; 'the paradox of self-amendment' in constitutional law; *tu quoque*, or 'equity must come with clean hands'; *renvoi* in conflict of laws; 'ignorance is no excuse'; . . . the 'fiction theory' of the legal person, according to which

> the State as a legal person must, like Munchhausen, pull itself up by its
> own bootstraps by inventing itself.
> Teubner (1993, pp. 4–5)

Such self-referential paradoxes are not, of course, confined to the law, but appear in other disciplines such as psychology and philosophy (Hughes and Brecht, 1978). There are various possible responses to the problem, none of them particularly satisfactory. One easy way out is to suggest that it is best to simply ignore them as being trivial puzzles of no real importance. Another is to argue that they are merely a linguistic problem—they come about merely in our language because of the different distinctions and categorizations that we make. The remedy is thus to tidy up our language by creating particular constructs or distinctions to bring back consistency such as, for example, Bertrand Russell's Theory of Logical Types. A third response, more specific to law, is the radical critique of the Critical Legal Studies movement (Joerges and Trubeck, 1989). This is a form of deconstruction which actually sets out to show that all aspects of the law are in fact full of contradictions and antinomies.

All these approaches are similar in treating self-reference and its paradoxes as something problematic to be avoided. In contrast, autopoietic theory treats these phenomena as central and constitutive of real systems. The law is full of circularity and self-reference, but rather than being avoided or banned they should be accepted and explored as key mechanisms in the establishment and development of the law.

9.2.   Autopoietic Law

We have already seen (in Chapter 8) Luhmann's overall approach to theorizing society as an autopoietic communicative system and the place of law within it. I shall briefly recall this before concentrating on law in particular. Luhmann defines society as consisting of a network of communications which continually regenerates itself. Communications trigger new communications in a continuous, self-defining process. A communication is not a communicative act nor a particular message but the connection of an utterance conveying information with the understanding it generates. Society consists only of communications and of all communications. Thus the physical world and even people and consciousness are all external to society. Communications are not entities but events produced at a point in time; they have a duration, but only that of their actual occurrence. They do not exist after their occurrence. When they are completed, their very absence calls for another event or else the system can no longer exist. Moreover, the next event cannot be a repetition of the first, but must be a different event.

Modern societies are not homogeneous, but have become internally differentiated. Particular types of communications (e.g., legal and economic ones) have separated themselves out from the rest to form their own self-defined autopoietic subsystems. These are not outside of or different from society; they are merely particular communicational domains that have achieved a certain autonomy from the others. Each subsystem becomes self-defined, operating around a particular distinction—in the case of the law, that of legal/illegal.

Thus Luhmann has a very radical conception of the legal system. It contains no laws or courts or parliaments or judges or lawyers or even defendants as such; it is pure communication:

> The legal system . . . consists only of communicative actions which engender legal consequences—it does not, for example, consist of physical events nor of isolated individual behavior. . . . It consists solely of the thematization of these and other events in a communication which treats them as legally relevant and thereby assigns itself to the legal system.
> Luhmann (1987b, p. 19)

So, what are legal communications and how do they differ from other communications? Luhmann tells us, rather unhelpfully, that a legal communication is one that has legal consequences; in typical fashion, he gives almost no actual examples at all. Some that come to mind are a registration of birth or death, the decision of a jury, the proclamation of a new law, the judgment of a court, the payment of a fine, the issuing of a parking ticket, and so on. The first point is that what counts as a legal communication will be defined by the system itself. This is in its very nature as an autopoietic system. The legal system is any and all legal communications, and its self-definition is its way of differentiating itself into an autonomous unity. But if we as outside observers cannot specify this in particular, we can at least indicate in general how it operates.

9.2.1. Normative and Cognitive Expectations

As mentioned in Chapter 8, communications are an integration of utterance, information, and understanding, and each of these is a selection from a "horizon" of possibilities. In order to cope with the complexity of the world, such selections are often made on the basis of "habit or expectation" (Luhmann, 1985a, p. 31ff). We do not consider each situation anew in depth but work on the basis of expectations established through the history of our interactions (see also Maturana on living as a cognitive activity, Section 5.2). Such expectations will inevitably be disappointed, and this may lead to us changing them (and learning), or we may maintain them despite the disappointment. This is especially the case with law, which can be seen as embodying a complex set of expectations or norms about what should or might happen. It is through their interaction with legal expectations that actions and events can be incorporated into legal communications.

Luhmann distinguishes two different kinds of legal expectation, *normative* and *cognitive*. Normative expectations are embodied in laws and concern what ought to happen—people ought to pay taxes; they ought not to steal. Such expectations are general and do not change in the light of disappointments. The fact that someone does not pay taxes or does steal does not change the norm. In fact, normative expectations are in a way precisely defined by such occurrences; they are "counterfactually stabilized" (Luhmann, 1985a, p. 33). They function because they mark boundaries that may be and are transgressed. In contrast, cognitive expectations concern what will happen: A. N. Other will pay her taxes or will not steal. These are particular expectations and are changed in response to disappointments. We are prepared at the outset for them not to be met, and we learn and change our expectations when they are not. Distinguishing between normative and cognitive expectations is not a semantic question but concerns our response to disappointment. If we are prepared to change the expectation, it is cognitive; if not, it is normative.

Luhmann uses this distinction as the basis for his characterization of law, for he claims that law is normatively closed but cognitively open. This is the mechanism by which he reconciles the closure of autopoiesis with the obvious interaction between the law and its environment.

If normative expectations cannot be changed by events, how can they be changed? The answer Luhmann gives is that they can be changed only internally, by the legal system itself. In this sense the legal system is a closed system:

> The autopoiesis of the legal system is normatively closed in that only the legal system can bestow legally normative quality on its elements and thereby constitute them as elements. Normativity has no purpose beyond this (in the sense of an intendable end). Its function is continuous making possible of self, from moment to moment, from event to event, from case to case. . . .
> Luhmann (1987b, p. 20)

This is the essence of the way Luhmann depicts the closed nature of the legal system, which seems to me to work in two different ways: first, it is only in this way that legality (i.e., having a legal aspect) is conferred on external components so that they become part of the legal system, and second, such norms can be determined only by the legal system itself.

Luhmann is saying that events and happenings can become the subject of legal communications only to the extent that they are covered by or subject to legal norms. The norms specify the conditions that must apply for something to become legally relevant. For example, at birth a child acquires particular legal rights. These are normative expectations about what should happen to a child, and it is only through the possibility of meeting or disappointing these expectations that births can be said to have legal relevance. What these expectations are, however, is entirely a matter for the law.

On the other hand, that a particular birth has occurred, or whether it had occurred at a particular time, or whether a particular child has been harmed are all cognitive matters. They are contingent facts, which may or may not turn out to be the case. This is how the system is open, and this is the source of new legal communications. A new birth does not acquire legality until the appropriate communication (registration of birth) has occurred, and this communication (in time) generates more communications. Always, however, the legal communication concerns only the legal aspects of the situation defined by the existing legal norms. Other aspects of the situation generate other communications, e.g., economic, political, religious, or simply lifeworld ones.

There are, of course, a multitude of nonlegal norms and expectations—social, aesthetic, political—which also apply to events and actions. These are essentially irrelevant to the legal system. The legal system does not import normative expectations, nor does it export them, nor does it refer to them in its operations; this is its organizational closure. Such norms can affect the legal system only if they can become converted into legal norms. For example, changing societal expectations about ethical business behavior may generate political communications leading to the enactment of a new law, which would then be a legal communication altering legal norms. The interaction between the law and other functional subsystems will be discussed below.

9.2.2.   The Binary Code: Legal/Illegal

It is one of the major achievements and operations of the autopoietic legal system that it does distinguish its own communications and norms from nonlegal ones. One way in which legal norms are distinguished from others is through the binary code of legal/illegal. [Luhmann (1992b) argues that all the functional subsystems revolve around particular binary codes, for example, truth/falsity (science), power/nonpower (politics)]. This is part of the mechanism of legal autopoiesis; all events and states about which legal communication can occur must ultimately be categorized as legal or illegal, and this is what drives the system (Luhmann, 1992a).

> The structure that actually organizes the autopoiesis of the system as an unavoidable outcome of its own operations is the system's binary code; that is, the continuous necessity of deciding between legal right and wrong. This code is a strictly internal structure. To declare something illegal does not mean that it belongs to the environment of the system.
> Luhmann (1992b, p. 1427)

There is nothing magical about the code, but it is ultimately that which characterizes legal communications. Questions concerning the legality or illegality of events or acts are thereby legal communications. Questions which are not concerned with legality are not legal communications:

> The question of legality is or is not picked up in communication, and by this very fact the communication takes part in the recursive network of legal communications.
> Luhmann (1992b, p. 1428)

This code is something fundamental to the system. It is not, itself, one of the system's norms but is at a metalevel. It is that to which all norms refer in the sense that they all define the boundary between legal and illegal. Norms (i.e., laws) can and do change, but the demarcation of illegality is the whole purpose of the legal system. Without it there would be no legal system. If the binary code itself were to be normative, then the paradox mentioned in Section 9.1.1 would apply, but the split between code and norm avoids this. It is also this distancing that allows for the positivity of law. The code remains the same, but the programs (the term Luhmann uses in *Ecological Communication*), which contain the rules for applying the code to particular events, can change. The term "rules" can be used here because Luhmann suggests that norms can often be expressed in a rule-like "if–then" formulation—"if taking the item was deliberate (cognitive condition), then the act was theft (normative conclusion)."

Note that this code, and therefore the legal system, does not have a moral or ethical dimension. It is not the same as right/wrong or good/bad. The law is concerned only with the specification of legality, as determined by its own self-referential communications. Morals and ethics are nonlegal norms enacted in other parts of society.

9.2.3.   The Legal System and Its Relation to Other Systems: Closure and Structural Coupling

Once a legal communication is generated, further communications refer only to legal terms and decisions, for example, "the plaintiff," "companies as defined in Section so-and-so," "the case of X," not to the original external factors. It is in this way that Luhmann can maintain that the legal system consists only of communications, not people or institutions. It does not matter which particular person (judge, juror, defendant) or institution (legislature, grand jury, court) initiates a particular communication or, indeed, receives and understands it. It needs only to be someone appropriate, as defined by the normative expectations of law. Whether a particular person is appropriate might of course be a matter of dispute, and thereby the subject of further legal communications. These, however, would never "escape" the legal system until, ultimately, an event is triggered which has extralegal consequences—for example, imprisoning someone would have familial and economic consequences.

However, this normative closure of the legal system does not mean that it is isolated or interactively closed. It is cognitively open and has particular interactions with a variety of other systems. These other entities, whether they are people (what Luhmann would call *psychic systems*) or institutions or other functional subsystems or society itself are in the environment of Luhmann's legal system. They can interact with it, in the sense that they can perturb it or irritate it, but they are not part of it, nor can they enter it—they can only trigger communications. In his early work, Luhmann's theory left these relationships underdeveloped and could be criticized (Mingers, 1991) for not using Maturana's important concept of structural coupling (see Section 3.1.4), but this has now been rectified (Luhmann, 1992b).

We have seen that the legal system is open through its cognitive expectations. Indeed, if it were completely closed or isolated it would cease to operate; it needs continual stimulation in the form of events requiring legal debate and classification. The normative and cognitive expectations presuppose that particular perturbations occur in the environment in the same way that the biological cell presupposes recurrent interactions, and through this the legal system becomes structurally coupled to the various environmental systems—consciousness, society as a whole, and other subsystems. Note that these perturbations are not determined as such in the environment. They are not intrinsically labelled "legal perturbation." Rather, the particular legal structure (of expectations) determines what can and cannot be a perturbation for it. For example, if there is no legal norm about a particular form of pollution, then it does not exist as far as the legal system is concerned, no matter how much it actually occurs.

The structural coupling to society and consciousness occurs in a general way, while that to other subsystems in a specific way. With regard to consciousness, communication is not, as we have seen, either thoughts or speech acts, but these are a necessary trigger for communication. And in the opposite direction, societal communication is a necessary perturbation for individual consciousnesses without which they would not become social beings. As for society, society itself is communication, and the legal system is just a specialized form. Societal (lifeworld) communication, e.g., a dispute or argument, can, under certain conditions determined by the legal system, trigger a legal communication, and this ultimately generates an effect on the lifeworld. There will be a reciprocity between the two determined not functionally but historically. The structures of expectation of the legal system are relevant to events that occur in society, or else they would never be triggered, although there will always remain obsolete norms no longer relevant but not removed.

There is more specific coupling with other functional subsystems. This can happen in two ways. First, there are particular communications that are actually shared by different subsystems; for example, the payment of a court fine is both legal and economic, the enactment of a law is both legal and political, and the establishment of a patent is legal, economic, and scientific. Such communications trigger further and quite separate communications in all relevant subsystems. More generally, particular realms of law are oriented specifically to another subsystem. For example, with regard to economics there are the laws of property, contract, corporation, and patent. These do not represent integrations of these subsystems, but structures that are mutually and reciprocally compatible.

> Modern concepts of property and contract do not integrate or even de-differentiate the legal and economic systems. As mechanisms of structural coupling, they organize the reciprocal irritation of these systems and influence, in the long run, the natural drift of structural developments in both systems.
> Luhmann (1992b, p. 1436)

9.3. Criticisms of Legal Autopoiesis

Needless to say, such a radical legal theory has generated much debate. Some of the criticism is similar to that concerning Luhmann's general theory (Section 8.5.4), and some is specific to law. I will group the main criticisms using categories similar to those of Kennealy (1987): first, those concerned with the status of the theory—whether it is a genuine theory or merely metaphorical or tautological; second, those concerned with the detail of the theory as an explanation of law; and third, those that consider the implications of the theory for law's autonomy and interaction. King (1993) has recently defended autopoietic law against a number of the criticisms that I discuss below.

9.3.1. The Status of Autopoietic Theory

The first, most basic, questions concern the status of autopoiesis as a theory. Does it actually make the strong ontological claims that legal systems really are autopoietic, or is it better seen simply as a sophisticated biological metaphor (Kennealy, 1987; Diamond, 1992; Zolo, 1992)? Alternatively, is it actually tautological, developing its own formal abstractions in a way that could never be put to an empirical test (Rottleuthner, 1987; Grzegorczyk, 1991; Munch, 1992; Wolfe, 1992)?

The "autopoiesis as a metaphor" view is certainly tempting, as it removes many of the problems of interpreting autopoiesis in a realist manner. At first sight it seems highly unlikely that a theory developed for biological cells should also be able to explain something as radically different as the legal system, so seeing it as a metaphor is much more acceptable. We have already seen (in Chapter 8) how Morgan uses autopoiesis as a metaphorical way of interpreting and understanding Organizations, and such a view might be quite revealing for the legal system. However, it is clear that the main proponents, Luhmann and Teubner, certainly do not see it as merely a metaphor, and to see it as such would almost entirely diminish its significance. I would suggest that it is necessary to take the strong claims first and see whether or not it is possible to sustain them.

The next claim is that the theory is formulated in such abstract terms that it could not be tested by any empirical evidence—that essentially it is a matter of definition and is therefore a tautology:

> . . . autopoietic theory cannot, in fact, be regarded as a description of social reality, since its object—the legal system—has no real existence, being merely an object of formal analysis, voluntarily created by a stipulative and hence arbitrary definition. . . . It is impossible to conceive of relational acts of pure communication as existing, other than by a formal abstraction.
> Grzegorczyk (1991, p. 130)

The argument here is that the theory is so abstract and analytical that it cannot describe actual reality. In excluding actors, motives, institutions, Organizations, and so on and in defining a system as pure communications, the theory has no empirical content. It is compatible with any state of the world and produces no hypotheses for consideration. Munch (1992) also argues that the theory's self-referential nature leads it to be contradictory.

It is certainly true that Luhmann's writing remains resolutely theoretical, with very little by way of empirical support, but overall the criticism seems to me to be misplaced. It may to some extent reflect a lack of understanding of autopoiesis—for example, Munch's comments on structural coupling (Munch, 1992, p. 1468), which he claims is Luhmann's invention and destroys the whole notion of autopoiesis, are clearly quite wrong. We have seen how autopoiesis has been used to explain a whole range of characteristics of the legal system that are otherwise difficult to explain. We may debate about how successful it is with respect to competing explanations, but it does appear to make empirically testable claims. Luhmann responds to this criticism (1992b, p. 1438) by saying that current empirical methods are not suitable for describing such complex, self-referential systems as law. Their use would tend to obscure rather than reveal autopoiesis. However, he sees no theoretical problem with empirical verification and makes a number of suggestions of possible hypotheses.

9.3.2.   Does Autopoiesis Provide a Valid Explanation of Law?

As for the substantive content of the theory, there are a whole range of issues, the main ones being

1. the question of boundaries (Kennealy, 1987; Diamond, 1992),
2. whether the notion of "production" is appropriate in this situation (Rottleuthner, 1987; Teubner, 1993, p. 20),
3. the focus on pure communication to the exclusion of all else (Kennealy, 1987), and
4. whether it is possible to omit people (Jacobson, 1989; Van Zandt, 1992; Wolfe, 1992).

The first two issues are general ones concerning Luhmann's theory as a whole, already covered in Section 8.5.4. To what extent is it a valid application of autopoiesis as originally specified by Maturana and Varela? As we have seen, the definition of autopoiesis specifies processes of production, within a boundary, producing both the boundary and the non-boundary constituents. With the legal system there is no boundary

as such although there is distinction between legal and nonlegal components, and it is not clear that the relationship between communications is one of production. On the boundary question, there is no doubt
that there is not an actual boundary demarcating the legal system, but
Luhmann does make a strong case, if one accepts communications as
the basic constituent, that legal communications can be clearly distinguished from others and that they form a self-referentially closed system.

I think that on the question of the production of communications
the theory is less satisfactory. Luhmann defines the notion of communication as including both that which is transmitted and that which is
understood, and he abstracts it from both sender and receiver. It then
becomes highly debatable whether it is reasonable to say that one communication produces another. It may stimulate or trigger it, but surely
it is the sender who actually produces the communication. To take a
trivial example, in the interchange "How are you?" "Very well, thanks,"
I think that it is stretching the notion of production too far to suggest
that the first produced the second. We need to distinguish between the
stimulation of one communication by another and the *production* of a
communication by an agent (Mayntz, 1987, quoted in Teubner, 1992).

The term *production* means "bringing into existence," and it is clear
that one communication cannot, of itself, bring another one into existence—this can only be done by an agent. The agent is necessary to
make the selection that constitutes the understanding of the received
message and the selections that constitute the information and utterance
of the next message and actually generate it. Luhmann himself recognizes the necessity of agents: "communication always presupposes an
addressee who is independent and can either accept or refuse the proposed meaning" (Krull *et al.*, 1989, p. 92). The addressee, of course,
becomes the sender of the next communication.

Varela also argues that the notions of production and boundaries
are not suitable for the social domain (as noted in Section 8.3), and I
would follow him in this. The implication of this, already discussed in
the general discussion of Luhmann (Section 8.5.4), is that the theory is
better described as one of organizational closure rather than autopoiesis.
Organizational closure simply requires that all changes of state in a system lead to further changes of state of the same type within it. Thus
legal communications lead to further legal communications. This would
maintain most of the detail of the theory without requiring the identifi-
cation of boundaries and the production of components.

The third problem identified was the restriction of the system to
pure communications, and the fourth was that people, in particular,

could not sensibly be excluded. To many (Kennealy, 1987), the legal system clearly includes legal institutions, such as courts, laws, crimes, and various legal actors such as judges, juries, criminals, and so on. A theory that has nothing to say about these and actually excludes them from its definition of a system cannot really be an adequate theory of law.

The omission of actors and their motives is what most critics object to, and it is also one of the general criticisms (Section 8.5.4). Wolfe (1992) identifies what he considers to be a central dilemma in Luhmann's theory. Luhmann is right to move away from the sociology of individualism or consciousness toward a structural or systems theory, but in so doing runs the danger of becoming so abstract that the theory is totally divorced from actual human activity.*

For Jacobson, it is the needs and desires of individuals that provide the necessary motivation for continually applying and changing the law:

> Luhmann misses the method and motor of common law self-generativity: the role of individuals—ordinary legal persons—in generating legal norms, and the need of individuals to keep transforming them. Jacobson (1989, p. 1681)

People have both a moral desire for rightness and a social desire for stable communities, and it is these which provide the dynamism of law. Without this, and given the criticism of communication producing communication outlined above, it is difficult to see why Luhmann's system should keep going.

Van Zandt (1992) makes a similar argument concerning the dynamism of law but goes further in putting forward an interactionist case that theories about law as a separate system are quite unnecessary. Society, and societal institutions, are simply the result of the situated activities and decisions of innumerable individual actors. These generate outcomes that cannot be seen as intended or intentional and place constraints on further action. These may appear to an individual as some entity or system in its own right, for example, "the market" or "the law," but in reality there is no such thing. There is just the ongoing activity and its effects on later activity. (For a more sophisticated analysis of the interaction between structure and action see Section 8.4.1, on Giddens.) Thus Luhmann attempts to explain something that does not exist any-

---

*The theoretical work of Habermas (1979, 1984, 1987) is aimed at precisely this point— how to analyze social action at a level beyond the individual and consciousness. His theory is based on *communicative action* rather than communication. Teubner (1989) presents an interesting comparison of Luhmann, Habermas, and Foucault.

way—a reification—and the dynamism of law is simply a result of the interaction of self-interested individuals.

9.3.3. *Autonomy and the Interactions of Law*

The autonomy of law and its interactions with the rest of society are two clearly related questions, but there are a number of different concerns. Considering first autonomy, there is disagreement about the actual notion of autonomy itself (Lempert, 1987; Nelken, 1987; Luhmann, 1987c), and Zolo (1992) argues that autopoietic autonomy is tautological since it is true by definition. The traditional, nonautopoietic view of the autonomy of law sees it as a characteristic of the relationship between law and particular interests within society. The law is seen as open to society and subject to possible influence and control; autonomy is seen as a continuum measuring the relative independence of law. It is, therefore, a normative concept—law should strive to be autonomous and to the extent that it is influenced and changed by society there is a loss of autonomy.

For autopoietic theory, on the other hand, autonomy is all or nothing; it is an inevitable feature of an autopoietic system. The law is a closed, self-defined system—only law makes law—so it is not open to determination by its environment. It is constantly changing, but these changes occur through its own structure and activity, such changes cannot be identified as responses to particular environmental pressures. The question is not how to maintain or improve the autonomy of law with respect to societal encroachment, but how an autonomous, organizationally closed system comes to have a structure that allows its continuation in a particular environment. The autopoietic answer to this is the historical drift of an autopoietic system while being structurally coupled to its environment. This leads us into the form of legal interaction.

The obverse of the question of autonomy is the nature of the interaction between law and the rest of society. As we have seen, the ideas of normative closure and autonomy do not make law noninteractive or isolated; it is cognitively open to events in its environment. But Luhmann's early writings were, I think justifiably, criticized (Kennealy, 1987) for a lack of attention to this area. This has been rectified with the more recent work on structural coupling and resonance (Luhmann, 1989b, 1992b). However, I do think that there are still some important debates about the extent to which law is nevertheless open to societal influence through the legal actors who enact it (Munch, 1992; Rosenfeld, 1992).

This debate is clearly related to that concerning the exclusion of people from the system. The basic argument can be put as follows: Even if we accept communications and normative closure as the proper theoretical characterization of the legal system, in practice extralegal norms and influences are brought in by people. Communications and decisions must ultimately be made and interpreted by particular agents who participate in society as a whole and so will, consciously or unconsciously, bring to bear a range of factors. Thus even though legal communications should in theory be based on legal norms, in practice they will be contaminated (i.e., constituted) by nonlegal expectations and influences.

It is easy to find examples throughout the legal system, from initial enforcement to final sentencing. In theory, the police should enforce all existing laws equally. In practice, because of finite resources, this is not possible and decisions are constantly made at all levels as to which legal norms to apply in particular situations. For example, the police in Britain typically ignore minor traffic offenses and (in recent years) only give cautions for possession of soft drugs and first-time shoplifting offenses. Yet they often invoke obscure and questionable laws to prevent pickets and demonstrators from traveling on the road. These decisions are made on political, financial, and moral grounds, not simply legal ones.

Once someone has been charged, it is up to the Crown Prosecution Service (in England) to decide whether or not a case should go to court (whether or not the next legal communication should order a trial). Again, there is considerable space for judgment in these decisions and the possibility of implicit or explicit nonlegal considerations playing their part. Once a case is brought to court it is clear that financial considerations affect the outcome—having the financial capacity to employ expensive legal counsel or even to defend oneself at all—and the existence of legal aid, itself politically controlled, does little to redress this. The decisions of juries will also not be purely legal. Jurors are ordinary people without legal training who inevitably bring with them their particular social backgrounds—class, race, gender, etc.—which will influence their decision. (Think of, for example, the Rodney King case in Los Angeles.) Finally, the same will be true of judges (who come from a very narrow segment of society). They have much influence in directing the jury and especially in sentencing, and it is clear that they exercise these powers in both an individualistic and a class-based manner.

In all these examples, it is not that the decisions made are against legal norms; they are generally all legal rather than illegal. Rather, it is that the framework of legal expectations is underdetermined with respect to the decisions. There is always considerable space for the exercise of choice and discretion, and this space becomes colonized by the structures and expectations of society as a whole. The legal system is therefore not as normatively closed as Luhmann argues.

9.4.   Conclusions

My conclusions about legal autopoiesis are much the same as those concerning Luhmann's general social theory. His work certainly provides a stimulating and very fruitful approach toward theorizing the law—witness the large amount of interest which it has generated. It also seems to address precisely those features, particularly evident in the law, such as its self-referentiality that other theoretical positions do not. And it provides a novel way of viewing the interaction between law and society, which can potentially do justice both to the autonomy of law and to its interactions with and coupling to other systems.

On the other hand, a number of related criticisms seem to me to have considerable force, in particular, the difficulties of using the autopoietic notions of production and boundaries, the reductive nature of the restriction of the legal system to pure communication, the inadequate analysis of the relationship between agents and communications, and the infiltration of the legal system by extralegal norms and expectations. It seems to me not impossible to deal with most of these points using a weaker organizational-closure version of the theory and developing a richer model of the way that agents and communications on the one hand and the legal system and society on the other interact.

Family Therapy

10.1.   Introduction

Psychoanalysis and psychotherapy include myriad different approaches, one of which is family therapy. The latter is characterized by therapists who work with families rather than individuals, seeing the symptoms manifested by the individual as related to the interactions of the family as a system. This approach stems from a basic systems perspective and has developed to reflect changes within the systems movement itself.

Section 10.2 looks at the historical emergence of a constructivist perspective in family therapy. Section 10.3 explains this position, and Section 10.4 discusses criticisms of it.

10.2.   Historical Developments in Family Therapy

Family therapy has always been strongly influenced by systems theory. Early work (e.g., Jackson, 1957) was based on first-order cybernetics (especially the work of Weiner, 1948), viewing the family as a self-stabilizing system and using ideas such as feedback and homeostasis. The root metaphor saw the family as a smoothly functioning machine with a hierarchical structure of roles and behaviors and the therapist as a mechanic able to repair such systems (Hoffman, 1988). The focus was on the maintenance and stability of the family organization. During the sixties this was further developed to focus on change in the family following the "second cybernetics" (not to be confused with "second-order cybernetics") work on positive feedback and deviation-amplifying processes (Maruyama, 1963; Buckley, 1968). The underlying epistemology was clearly objectivist—the family was an independent unit that could be objectively analyzed by a neutral analyst.

A major shift occurred during the seventies, based on the work of Gregory Bateson (1973, 1979), who can be seen as forming a bridge between the objectivism of first-order cybernetics and the constructivism of Maturana's second-order cybernetics. First, Bateson emphasized the importance of circular and reciprocal chains of mutual causality rather than linear thinking (such as A causes B). The family was therefore seen as a system of symmetrical and complementary behaviors and interactions, and notions such as hierarchy, and power and control were abandoned. Second, there was a shift away from the metaphors of energy and matter toward those of information, context, and meaning. Meaning was no longer objectively given but transformed and modified by context and transmission. Sequences of interactions were repetitive and circular and could be split up or punctuated in different ways by various members of the family. Even more radically, this approach pointed to the importance of the therapist as punctuator and definer of reality, as part of the system rather than an independent objective observer (Watzlawick, 1967).

This approach was most formalized in what is known as the Milan method developed by Selvini-Palazzoli *et al.* (1978; Campbell and Draper, 1985; Tomm, 1984). This can broadly be described by three terms: *hypothesizing, circularity,* and *neutrality.* Therapists generate *systemic hypotheses* involving circular relations among all family members. These are then investigated by questioning individuals about the differences and relationships between other family members. The therapist strives to remain neutral—that is, not to side with any individual in the family. Interestingly, this group of therapists split up during the eighties and are now pursuing two different directions (Jones, 1988). Boscolo and Cecchin have been strongly influenced by Maturana and have taken up a constructivist position of almost nonintervention, whereas Selvini-Palazzoli and Prata have moved in the opposite direction, seeking objective truth and, if necessary, taking an adversarial stance toward families.

The Batesonian approach itself was then superseded by the more radical ideas of Maturana and Varela, under the rubric of a constructivist or "bringing forth" paradigm (Birleson, 1988). Superseded is not quite the right word—as we shall see, Maturana's work has been as contentious in family therapy as it has been in other disciplines and is the subject of much debate. Other recent trends are, first, that some family therapists have widened their focus beyond the family to be willing to deal, at least initially, with individuals within families and systems wider than the family from a systemic viewpoint (Wynne *et al.*, 1986; Jenkins and Asen, 1992; Jones, 1993). And, second, that there has been a strong development of feminist family therapy, which has had an uneasy relationship with systemic therapy (Goldner, 1991).

10.3.   Constructivist Family Therapy

Maturana's ideas were brought into family therapy in the early eighties by Dell (1982a, 1982b, 1985), Watzlawick (1984), and Keeney (1982, 1983). There is a good introduction in a special issue of *Family Therapy Networker* (Simon, 1985a, 1985b; Efran and Lukens, 1985), some spirited debate in a special issue of the *Irish Journal of Psychology* (Kenny and Gardner, 1988; Goolishian and Winderman, 1988; Hoffman, 1988; Mendez *et al.*, 1988), and other introductory articles by Leyland (1988), Efran *et al.* (1988), Ludewig (1989), Griffith *et al.* (1990), and Varela (1989). Gergen (1985) argues for a constructivist position in psychology in general, Delmonte (1989) has compared constructivism with existential psychotherapy, and Wright and Levac (1992) apply Maturana's ideas to noncompliance in nursing. It is mainly Maturana's later work that has been of interest in family therapy. As this is described in Sections 6.3 and 7.4 I will review it only briefly.

Maturana argues that each person creates and constructs his or her own world of experience through the distinctions he or she makes, or "brings forth," in language. Because of the structural coupling that develops between people and the consensual, intersubjective nature of language itself, these individual worlds intersect and overlap. But all of the realities so generated must be respected as equally valid, although not necessarily equally desirable. There can be no outside, privileged viewpoint such as that of a therapist. Equally, there is no independent, objective world to which we can appeal to validate our beliefs or constructions. Rather, we must "put objectivity in parentheses" and accept that there is a *multiverse* of different realities. Maturana calls this a *constitutive ontology,* according to which things are brought into being (constituted) by being distinguished, as opposed to a *transcendental ontology,* which assumes access to a true reality. Interaction between human beings takes place through *languaging*—a complex meshing of language, body, and emotion. We can describe such interaction through recurrent conversations in which we participate.

The family constitutes, for Maturana, a social system (see Section 8.4). That is, it forms the medium or environment in which the autopoiesis of a number of individuals is realized. It exists as a set of recurrent conversations between its members through their structural coupling. Such a system is highly conservative in that it reinforces behaviors that confirm it, and continued membership requires the maintenance of particular patterns. This is so even if one or more members dislike it. Each family is different, each family realizes its own particular pattern of interactions, and replays its own particular conversations.

Moreover, each member experiences this differently, experiences a different family from the others. All such experiences are equally valid.

One cannot interact with a family as such, only with its members. The organization of a particular family is constituted by the structures of its members and, therefore, change can come about only through change in the members' behavior. The fact that a family is a closed network of conversations has implications for outside intervention. Any interaction with an external entity (such as a therapist) acts merely as a perturbation for the family, which has to find its own compensation. This compensation depends on the operation of the family as a system, on the particular structures of its members at that time, and not on their intentions or the nature of the interaction. This in part explains the common psychological phenomenon of resistance.

What are the implications of this constructivist position for family therapy? First, families come for help when the realities produced by their particular conversations become sufficiently unpleasant or distressing for them. Maturana (Mendez *et al.*, 1988) suggests that this usually comes about because many conversations attempt to impose one person's characterizations and expectations on another rather than accepting the different worlds that each individually brings forth. Thus, there are conversations for *characterization* and conversations for *accusation* and *recrimination*. Characterization takes place when one person has expectations that have not been agreed upon about the other members' characters and future actions. This then leads, after the event, to unjust accusations and recriminations when the expectations are unfulfilled. Maturana also talks of conversations for *coordinations of action,* in which requests and promises are made openly and agreement sought, which lead to mutually acceptable interactions.

A family may generate a particular closed organization of mutually antagonistic conversations, with each member viewing his or her own situation as the objectively correct one:

> If what is brought forth is a network of conversations for mutual characterizations, accusations and recriminations that are deemed objective, what is brought forth is a family defined as a network of conversations that entail impossible demands . . . and it constitutes the organization of the particular family. . . . As such, that organization both realises and generates the existential contradiction of the particular family. . . .
> Mendez *et al.* (1988, p. 158)

Given such a situation, what now can be the role of the therapist? It is clear that it must be radically different from that suggested by first-order cybernetics by both recognizing or defining a "problem" and of taking action to alleviate it. Now there can be no single, identifiable problem for the therapist to define. With objectivity in parenthesis, all views, all constructions are equally valid. The individual family members each bring forth a particular reality, and so will the therapist, but the therapist's will have no greater claim to validity—it will simply be another construction.

This may sound similar to the Batesonian argument that reality can be punctuated in different ways by various people. However, it is much more radical in that it denies that there is a single independent reality to be punctuated. Rather, there are a whole set of different realities being constructed in parallel and, at times, intersecting and interacting. There are no objective problems, merely mismatches between different constructions and the expectations they engender.

What role is left for the therapist? Families come in distress and needing help, but what can be done? We must remember a number of points: the family members create their own organization, and outside interactions will only be perturbations—they will be compensated by the family in accordance with its own organization. The therapist cannot have "instructive" interactions in which the effect of an intervention is determined by the intention of the intervenor. The family organization is constituted by its own members' behavior, so the only way to change it is by changing the members. As well as this, the family organization is conservative of its current organization, so any interactions that the therapist has with members must be ones that do not confirm the current patterns.

Next, the therapist must become involved with the family, must become part of its conversations and discover the recurrent pattern of actions that characterize it. But, of course, the therapist can never be a member of the family, at least the family as it normally is, for families never remain the same if the components change. For example, a family with a new baby is a new family; it is not the same old family with a new member. A new "temporary family" may be created from the conversations that involve or are triggered by the therapist. The therapist can no longer be seen as some outside, independent change agent, but as someone who participates in the structural drift of the family.

Once involved, how is the therapist to bring about change? The only way is to disintegrate or destroy the family organization as it is and for the members to constitute a new one through different behavior. To accomplish this the therapist must interact with the family members, individually or as a group, so as to trigger new conversations, which can be the basis of a more acceptable (to the family) organization. The problem with this is finding interventions that do not confirm the present family organization and behavior patterns. Maturana calls such interactions *orthogonal*. The family is constituted by the behavior of its members, but its members have other, outside, interactions as well as their family interactions. It is these which Maturana calls orthogonal (to the family), while those interactions which are part of the family, and thereby confirm it, are nonorthogonal.

The therapist must aim for orthogonal interactions that will not maintain the family as it is but may lead to changes in the individuals so that they are no longer able to participate in the current family conversations. Quite what such interactions might be, of course, vary from family to family. Discovering successful ones depends very much on the therapist's involvement and participation in the family in order to understand its particular linguistic features. In short, the therapist must become structurally coupled with the family members. Ultimately, provided that there is enough commitment to living together, the family is recreated with an organization that is mutually agreeable and supporting.

10.4. Criticisms of the Constructivist Position

Unsurprisingly, there have been fierce criticisms of the constructivist perspective. These fall into two main areas. First, some attack Maturana's radical epistemology and particularly his argument that there is no independent reality. These criticisms are generally similar to those discussed in Section 7.6. Second, some argue that constructivism is limited by its subjectivism and cannot address questions of social structure and, particularly, of power.

10.4.1. Criticisms of Constructivist Epistemology

Held and Pols (1985a, 1985b) initiated this debate by pointing out that the term "epistemology" was, apparently unwittingly, being used in two different ways within family therapy, and, more significantly, that a constructivist epistemology was essentially contradictory. They further argued (Held and Pols, 1987a) that the use of Maturana's concepts of structural determinism by Dell (1985) did not avoid this contradiction. Dell (1987) replied to this, quoting Maturana extensively, provoking a further response by Held and Pols (1987b). The debate continues.

The general meaning of "epistemology" is the study of knowledge as discussed in Section 7.3.2.2. However, in family therapy the term was often used differently to refer to a particular view about the world. For example, in distinguishing between a view of causality as linear or circular, reference might be made to a "linear epistemology."

This habit stemmed from Bateson (1973, 1979) himself, who used the word *epistemology* extensively in a number of different senses. In fact, Dell (1985) outlines five different ways in which Bateson used the term—as a theory of knowledge (the standard sense); as connoting a particular view of the world (e.g., that causality is circular); as a biological cosmology (i.e., that biological systems "think" and "decide" and are therefore epistemic); as a science (epistemology is the study of how organisms think); and as a character structure (what personal epistemology someone may have). It is the first two of these that are confused in family therapy, where the term is used in both the general and the particular senses. Held and Pols (1985a) suggest that the second meaning is illegitimate, as it really concerns the nature of the world—how the world is—and is therefore a question of ontology (or metaphysics) rather than epistemology. In this section, epistemology will be used only in its standard sense.

However, of more importance than the terminological problems is the argument of Held and Pols that any constructivist epistemology is inevitably contradictory. (A similar argument is made in Section 7.6.2.) They characterize the general constructivist position as one that denies the possibility of an "independent reality [being] attainable by the knower" and contrast this with the view that "independent reality [is] sometimes attainable by the knower" (Held and Pols, 1985a, p. 513). In later papers they refer to these as *antirealist* and *realist* positions. They recognize that there is a range of possibilities within each. In particular, within constructivism one can distinguish between the claim that our experience is inevitably structured by ourselves and so may be different from reality; the claim that there is an independent reality, but that we can never know it at all; and the claim that there is no independent reality and all reality is created by ourselves.

Although these differ considerably in their degree of radicalness, Held and Pols argue that, if taken seriously, they all deny the possibility of true knowledge of an independent reality. It thus becomes contradictory to accept constructivism and also claim to have a theory of how the "world really is." Yet this is just what Maturana (and following on, Dell) does in putting forward his theories about human beings. Maturana is making reality claims about the true nature of people.

If, as Dell (1985) suggests, *"all* observations have equal validity" since "perception is not and never can be objective" (p. 16) then on *what ontological basis* (in the common sense) can he maintain that Maturana's struc-

ture-determined *view* of the universe is "an enormously comprehensive
and powerful set of distinctions" (p. 17)?
Held and Pols (1987a, p. 460, original emphases)

Dell (1987), in reply, accepts that he may not have himself been
clear and then simply reiterates Maturana's arguments. These are that
as observers we always operate within language, generating explanations of our experiences; that we must always be aware of our biological
constitution and the limits it places on us and, in particular, that all our
interactions are subject-dependent and that we cannot distinguish between illusion and perception; that there is no independent reality, but
rather reality is constituted by us within our languaging; and that therefore explanations, even scientific ones, do not refer to some independent reality, but simply explain our own experiences.

Thus, Dell argues, Maturana cannot be accused of making illicit
reality claims since he denies the existence of any such reality and denies
that his explanations could refer to it. This reply is not satisfactory to
Held and Pols (1987b) nor, as should be evident from Chapter 7, for me.
Maturana may believe he is making no reality claims, but, in reality, he
is by privileging his own theories over others. He is making both an
epistemological claim that reality is unknowable and an ontological
claim that all observers really do have certain (biological) characteristics.
Debate on this central issue has continued (Coyne, 1985; Cade, 1986;
Bogdan, 1988; Sluzki, 1988; Cottone, 1989; Held, 1990; Oz, 1991; Held,
1991) but I shall consider only two other critiques, those of Speed and
Birch.

Speed (1984, 1991) contrasts constructivist approaches to family
therapy with realist ones, but also relates this to constructivism in wider
fields such as psychology (Gergen, 1985) and sociology (Berger and
Luckmann, 1967), and goes on to advocate a position she calls "co-constructivism." Speed accepts the constructivist notions that people construct their own ideas and conversations but emphasizes the social
nature of this process (something with which Maturana would certainly
agree). Language and the conversations that occur in it are inherently
social and consensual. However, she sees a problem in explaining why
particular ideas and conversations occur and not others, why some become dominant and others do not. A similar point is made by Mingers
(1984) in a general critique of subjectivist theories. This, to some extent,
foreshadows the criticisms to be discussed in the next section.

Her answer, co-constructivism, assumes that a structured reality
does exist and that it partially determines what we know. While we do,
as individuals and more generally as members of groups, make our own

distinctions and thereby construct our descriptions, such choices are made, often unconsciously, because of structures outside ourselves. Our membership in particular social and professional groups, our position in society, and our membership in that society rather than some other, as well as our own individual life experiences, all lead us to construct particular descriptions and not others. This view seems reminiscent of Marx: "Men make their own history but they do not make it just as they please . . . but under circumstances directly encountered, given, and transmitted from the past" (Marx, 1935, p. 245).

Birch's (1991) paper, although published in the *Journal of Family Therapy*, is actually a thoroughgoing critique of the whole of Maturana's and Varela's work, particularly the basic autopoietic concepts, rather than the later ones used by family therapy. This provoked a detailed response from Maturana (1991a) and my general conclusions are that much of the criticism is misplaced, based on a misunderstanding of Maturana's theories.

10.4.2. Criticisms of Constructivist Subjectivism

In this section I will consider a set of criticisms of constructivism in particular, some of which also apply to systemic approaches in general. These criticisms concern an apparent inability, or at least unwillingness, to consider the social and political context of family therapy. There are broadly four, partly interrelated, criticisms. The main one is that there is either no recognition of the importance of power, or an objectionable understanding of it as complicity between the wielder and the sufferer. The other criticisms all appear within the debate about power and so will not be discussed separately. They are first, that the "multiverse" argument that all views are equally valid means that one must equally accept any form of behavior, even violence and murder. Second, the family tends to be seen as an autonomous entity, isolated from the wider society. All problems are seen as problems of mismatch within the family rather than reflections of conflicts and contradictions within society. Moreover, there is no recognition of the socially constructed nature of beliefs and behavior. Third, insufficient attention is paid to the problems of gender—the theory and practice are unthinkingly patriarchal.

The problem of power is not specific to Maturana but concerns systemic approaches as a whole (Dell, 1989; Hoffman, 1990; Fish, 1990; Mens-Verhulst, 1991; Jones, 1993). In fact, it goes back to Bateson (1973), who argued that power is an inappropriate concept to use for systems. For Bateson, the idea of power as the ability of one entity to impose

control over another is a myth. This is because systems always consist
of strongly connected components and causality is always multiple and
circular. No one part of the system can gain unilateral control over the
others; recurrent patterns of behavior always stem from mutual interactions between one or more components. He did recognize that people
might believe this myth but argued that it was a mistaken way of thinking (epistemology, for him), as it would always lead to a failure of the
system.

For the family, this means that the overall pattern of behavior can
be related only to the family as a whole. No one person can be seen as
having power over the others, and in any relationships which appear to
involve dominance and submission the two roles are complementary—
the submission is as much a "cause" of the dominance as *vice versa:*

> . . . the family is a cybernetic system . . . and usually when systemic pa-
> thology occurs, the members blame each other, or sometimes themselves.
> But the truth of the matter is that both these alternatives are fundamen-
> tally arrogant. Either alternative assumes that the human being has total
> power over the system of which he or she is part.
> Bateson, (1973, p. 413)

This has led to a reluctance on the part of family therapists to use
terms such as *abuser* and *abused*, as they would imply a linear causality.
Milan therapists in particular have avoided blaming one person as the
"cause" of a problem: "participants are caught in a recursive pattern . . .
more like a misfortune, calling for compassion for the persons involved
rather than condemnation" (Tomm, 1984, p. 118). Indeed, Dell (1989)
suggests that family therapy has been remarkably reluctant to discuss
violence at all—*Family Process* has published only half a dozen papers
on it in 25 years—and yet the extent of family violence and abuse is
becoming ever more clear. More recently, male violence is becoming the
subject of more overt research, particularly from a feminist perspective
(Goldner *et al.*, 1990).

Maturana shares a very similar view of power. His concept of structural determinism and the impossibility of instructive interactions
means that, for him, one entity can never totally determine the behavior
of another. This means that power can be exerted only to the extent that
there is submission or obedience: "power is a matter of obedience, and
. . . nobody possesses power but is given it in the obedience of those
who obey" (Krull *et al.*, 1989, p. 95). "Power is action through obedience. . . . We always concede power in order to conserve something,
company, things, prestige, appearances, life" (Krull *et al.*, 1988, p. 98).
He further links this to objectivity-in-parenthesis. People who do not

accept this but believe that they have a privileged access to reality use this to force obedience. Maturana also explicitly denies that the family should be characterized by power relations as such relations do not constitute the biological nature of the family: "if we describe the family . . . as a system defined as a network of relations of power, we do not bring forth a family as a system that exists in the biology of the consulting people. . . . We bring forth a literary entity (Mendez *et al.*, 1988, p. 162).

So what are the objections to this nonview of power? First, it should be said that not all systemic therapists accept it. Haley (1976), who worked with Bateson, always argued that power is a characteristic of human nature and therefore central to family and social life, and Selvini-Palazzoli (1986), one of the Milan therapists, now considers that families are full of power strategies and "dirty" games.

The first problem, particularly from a feminist perspective (James and McIntyre, 1983; Taggert, 1985; Goldner, 1985, 1988, 1991; MacKinnon and Miller, 1987), is that power relationships are seen to be essentially ones of equal responsibility between the dominant and the subservient party. Yet such interactions are not at all equal in how much influence each participant has. This is particularly apparent in cases such as child abuse, rape (rape in marriage in now legally recognized in Britain), and wife-battering but is also constitutive of the more routine, daily interactions of many people because of their gender, class, or race. Moreover, this systemic attitude often leads to the rape or violence being seen as due to the *victim's* characteristics or behavior—the victim often takes the blame and sees herself in that way. And, at a most basic level, a failure to clearly identify an aggressor may endanger someone's safety.

This problem occurs partly because of the argument that power can never be linear or unilateral and partly because of Maturana's belief that all views are equally valid (although not equally desirable, see Section 12.3.5). Dell (who earlier was seen as a supporter of Bateson and Maturana) has concluded that:

> the circular–causal systemic perspective inevitably must obscure the existence of lineal power or violence . . . this blindness occurs because the systemic view is constitutively unable either to distinguish, or to speak of, "power", "violence", "abusers," or "victims."
> Dell (1989, p. 10)

I would certainly agree and say that many cases of violence are precisely that—situations in which one party does have total control over another. But even in less extreme situations, we do not need to talk of "total control" to be able to recognize the exercise of power. Bateson, in fact, does not deny power relations but argues that they are pathological, an error of thinking, which should be avoided. The implication of this is that they can be avoided; however, many would argue that humans are, by nature, strategic animals (Dell, 1989, pp. 4ff). Even Maturana does not deny the existence of power relations, claiming, however, that they are not social relations (which for Maturana would be based on care and love) because they are based on obedience and that they are intrinsically unpleasant and unstable (Krull *et al.*, 1989, p. 98).

The second criticism is that both Bateson and Maturana are assuming a very simplistic and individualistic view of power, namely, one person consciously exerting power over another. (See, for example, Perelberg and Miller, 1990; Goldner, 1991.) Work in sociology has shown that power is a much more complex phenomenon (Mingers, 1992c). Power is not simply subjectivist (exercised by individuals) nor objectivist (exercised by societal structures) but relational. It is both coercive and oppressive (power over) and enabling and productive (power to); its effects are both visible and invisible; it can be both intentional and unintentional; and it must be analyzed from the perspective of both agency which does, and structure which enables. Note the two meanings of "to be able to"—to be allowed to, and to have the capacity to.

Two particularly interesting analyses of power are those of Foucault (1977, 1982) and Callon and Latour (Callon, 1986). Foucault argues that power is immanent throughout society, involved in every interaction and that it is inextricably connected with knowledge (with which Maturana would agree)—knowledge is power, and power shapes knowledge. Callon and Latour study the practical exercise of power in science and demonstrate empirically how structures of interest and power are generated and reproduced.

The third weakness of the systemic view of power is that it fails to notice that power relationships generally follow particular, socially created patterns—men exercise power over women, whites over blacks, the rich over the poor, adults over children. These in turn both reflect and reinforce the basic inequalities of society. So long as family therapy fails to acknowledge these, it will inevitably reinforce them:

> By defining power simplistically, dismissing it and offering no other description for inequity in relationships characterized by domination and exploitation, the new epistemology lies in danger of mystifying issues concerning inequality in social arrangements. These differences in "power" are not random or arbitrary, but reflect a person's position within the socioeconomic system, based to a large extent on gender, race and class.
> MacKinnon and Miller (1987, p. 147)

This is one aspect of the more general problem of seeing families as systems, but systems without a context or environment (James and McIntyre, 1983). This leads to the assumption that all family problems stem purely from the dynamics of the family itself and that they can thus be solved. The alternative view suggests that society actually creates paradoxical or contradictory demands on families through the images that it projects. For example, the enormous pressure for consumption—home ownership, furniture, cars, vacations—cannot possibly be realized by a significant proportion of poorly paid or unemployed families. Another example is the contradictory demands society makes on women to be both traditional homemakers and active breadwinners. This also has implications for the possibility of change. For constructivist therapy, change comes about by changing the individual's behavior and ideas; however, if the difficulties arise from the external structure within which the family exists, then such change will rarely be successful.

Indeed, it is possible to view the therapist as an agent of social control (MacKinnon and Miller, 1987), implicitly adopting society's definition of the nuclear family and its accompanying roles and attempting to bring about accommodations within the family that allow it to meet them.

10.5. Conclusions about Constructivist Family Therapy

Constructivist family therapy is based both on Maturana's general cognitive theories and epistemology and on his theory of social systems. I think that it does bring some benefits to family therapy, but, for myself, there are major weaknesses in both areas.

The benefits are that it focuses attention on the way in which we individually construct our own versions of reality; it emphasizes interactions in the family as a whole, the fact that different families construct different realities and the conservative and autonomous nature of the family organization; and it brings to our attention the therapist's construction of reality and need for inclusion in the system.

However, epistemologically I do not accept Maturana's conclusions that there is no independent reality accessible to us, and socially I accept the criticisms that constructivism incorrectly ignores social and political contexts.

11

Information Systems, Cognitive Science, and Artificial Intelligence

11.1. Introduction

This final chapter of applications will look at two related areas: computer-based information systems (IS), and cognitive science and artificial intelligence. Within IS, autopoietic ideas have so far had only a limited influence, largely through the work of Winograd and Flores, who have produced a critique of both information systems design and artificial intelligence. This is covered in Section 11.2. More significantly, Varela has concentrated, in recent years, on cognitive science and has developed a new framework, the *enactive* approach, based partly on earlier cognitive theories and the phenomenology of Merleau-Ponty. This is the subject of Section 11.3.

11.2. Information Systems

11.2.1. The Language/Action Approach of Winograd and Flores

Maturana's theories have been imported into the study of computer systems through the work of Flores and Winograd (Flores and Ludlow, 1981; Winograd and Flores, 1987a; Kensing and Winograd, 1991). In *Understanding Computers and Cognition*, Winograd and Flores assimilate the phenomenology and hermeneutics of Heidegger (1962) and Gadamer (1975), the theory of speech acts of Searle (1969), and Maturana's cognitive theories to produce a critique of the traditional objectivist, rationalist approach to computer systems design and artificial intelligence (AI). In its place, they suggest an approach based on *conversations* and *commitments*, which they generally refer to as the "language/action approach." However, this name is also used by a group of mainly Scandinavian writers (Goldkuhl and Lyytinen, 1982, 1984; Lyytinen and Klein, 1985; Lehtinen and Lyytinen, 1986; Lyytinen *et al.*, 1991), who do not base their ideas on Maturana but rather on Habermas (1979). There is, however, considerable similarity in the emphasis on language, language as action, and speech act theory.

The main outlines are, first, that cognition and thought is not an isolated, separate mental function but our normal everyday activity—our 'being-in-the-world.' It is embodied in the patterns of behavior triggered by our interactions, which have developed through our structural coupling. "Thinking" is not detached reflection but part of our basic attitude toward the world—one of continual purposeful action. Second, knowledge does not consist of representations, in individuals' heads, of objective independent entities. Rather, we make distinctions through our language in the course of our interactions with others, continually structuring and restructuring the world as we coordinate our purposeful activities. Third, what is said does not occur *de novo*, but is grounded in our past experiences and tradition—the history of our structural couplings.

Fourth, the most important dimension of our actions as humans is language, but we must change our view of language from seeing it as representational and denotative toward seeing it as (social) action, through which we coordinate our activity. Languaging takes place in conversations which become the central unit of analysis. Such conversations are networks of distinctions, requests, and commitments, valid in respect of their acceptance by others rather than their correspondence to an external reality. Finally, the view of "problems" that computers can help "solve" must change. Problems are not objective features of the world, but the result of breakdowns within our structural coupling to objects or to others. When our activities do not succeed or our coordinations fail, our routine operation is disrupted and a "problem" occurs. This is always against a particular background and for a particular individual or group, and the nature of the problem becomes defined only through the attempts to repair it.

These ideas lead to a distinctive view about both the development of information systems in Organizations and the nature of computers and artificial intelligence (Smith, 1991). Organizations are seen as networks of recurrent and recursive conversations between individuals and groups of individuals (cf. the family—Chapter 10). The conversations consist of speech acts mainly involving requests, promises, commitments, and declarations coordinating general activities and the conversations themselves. Information systems should be designed to be part of and facilitate this communicative and coordinating process. They must be open and flexible, reflecting the changing distinctions and conversations generated within a domain rather than imposing an external and unchanging strait-jacket. A piece of software embodying these principles (called "The Coordinator") has been developed (Winograd, 1987; Flores *et al.*, 1988).

Equally, Winograd and Flores suggest that the whole objectivist thrust of computing/AI is misdirected. Developing systems to perform more and more complex calculations or better processing of chunks of reified "information" or "knowledge" will not lead to more human-like cognitive abilities. For this, one would need something radically different—a system, capable of significant structural change, which was able to develop its own readinesses and distinctions through a history of interactions in a domain that was of significance for its own operation.

The reception of this work has been highly polarized, as might be expected. Shortly after publication, four reviews were published in *Artificial Intelligence* (together with a reply—Winograd and Flores, 1987b), and these cover a broad spectrum of responses. Vellino (1987) writes as a philosopher, criticizing their characterization of the "rationalistic" tradition of AI and computing and rejecting their criticisms of it. He argues that rationalism has a perfectly adequate language with which to discuss its own limitations and that there is no need to adopt any kind of phenomenology; that analytic philosophy has at least begun to deal with the problems of the context-dependence of meaning and that while it is a difficult problem it has not been shown to be impossible; and that they simplistically misrepresent the nature of science.

Stefik and Bobrow (1987) are main-line AI researchers in the classic tradition. They mount a defense by arguing that the criticisms made by Winograd and Flores may apply to current AI software and algorithms, but that this does not prove the impossibility of the traditional approach and that in the future such limitations will be overcome. In general, they deny the whole basis of Winograd and Flores' argument, asserting that symbolic representation will eventually be proved to be sufficient. I am not sure that they would have so much confidence if they were writing now (1994)—see Section 11.3.1. Suchman (1987), with an anthropological background, is much more sympathetic. She accepts the basic critique of rationalism—although noting that, necessarily, any such delineation must be something of a caricature—and the thrust of their language/action view, although she does have reservations about their method. In particular, she argues that the use of Searle's speech act theory, and the particular "Coordinator" software, go against the basic position in being much too explicit and rationalistic. She also points to the lack of a strong social theory in Winograd and Flores' work and suggests parallels, particularly with ethnomethodology.

Finally, Clancey (1987) was most enthusiastic about the book. He came from a background of practically developing expert systems and found that it was very much in accord with his own experiences, but provided for him an important source of new insights and ideas and, indeed, changed his way of thinking. He was not wholly uncritical, however, arguing that there was a role for representationalism when analyzing reflective thought—i.e., when there was conscious reflection before action. Winograd and Flores' (1987b) reaction to the diverse reviews as a whole was to claim that they actually validated the approach. Each reviewer had brought his or her own particular background and prejudices and interpreted the work in his or her own particular way. Their work has certainly generated interest in a wide range of disciplines (Steier, 1987; Ducret, 1987; Strong, 1988; Bowers, 1991).

11.2.2. *Other Autopoietically Based IS Work*

Although Winograd and Flores' book is well known, there has been little substantive work in IS developed from autopoiesis. Stephens and Wood (1991) presented a general description of a constructivist approach to information systems, and Harnden (1990; Harnden and Mullery, 1991) has outlined ideas for what he calls enabling network systems (ENS). Harnden and Stringer are currently incorporating some of these ideas into an architecture for multimedia systems and the design of hypermedia (Harnden and Stringer, 1993a, 1993b; Stringer, 1992).

Harnden's work is based on the recognition that there is an enormous divide between the amorphousness of the way people think and the determinateness and precision of computer information systems (and other regulated systems of knowledge). As we have seen in Chapters 5 and 6, human thought is essentially linguistic behavior (languaging) and, as such, is context bound and consensual. We understand one another (to the extent that we do) not because language conveys definite information *per se*, but because it is a system of agreed symbols, which trigger similar reactions among people for whom it is meaningful. Thus information and communication occur among groups of people who are structurally coupled. Even so, individuals construct their own meanings within conversations, depending on their particular structural state at the time. Any communication or information is thus always dependent on context—the context of particular groups or individuals who interpret in particular ways, or may not be able to interpret at all.

In contrast to this, information systems are rigid and closed. Each one has its own limited set of syntactic and semantic rules and assumes that users understand and obey them. The aim of an ENS is to provide an interface between these two very different domains. Its functions would be twofold—first, to provide access to many different (and usually incompatible) information sources (public and private databases, bulletin boards, etc.) and second, to provide a space for users to explore their own and one another's cognitive domains. The latter is the more interesting. Users, in interacting with the system, will build their own personal domain of constructs and associations based on their own personal context and history. However, the aim is to be able to link these local consensual domains through some basic translation mechanism, and particularly to link them to the available information sources. Thus users with their own particular biases or prejudices should be enabled to access relevant information in an entirely flexible and natural way.

With current advances in multimedia production, Harnden and Stringer are focusing on the power of tactile interactive objects that engage a learner in a preanalytical manner, encouraging readiness for interest, motivation, and enjoyment.

Mingers (1993a,b) has used Maturana's ideas, together with work by Habermas (1979) and Dretske (1981), to conceptualize the basic nature of information. As explained above, information cannot be pure objective fact ready to be processed in information systems. Rather, information is very much dependent on the meaning structure or cognitive domain of the originator and receiver. The problem is then that of distinguishing between *information* and *meaning*. Does a sentence carry information, which then generates meaning in the receiver, or does it have a meaning, which may provide information for the receiver? Dretske chooses the former in creating a theory of how information in the world triggers meaning and knowledge in people. Mingers combines this with Habermas' work on *communicative action* and *universal pragmatics* to provide a framework covering both semantic and pragmatic aspects of information.

11.3. *Cognitive Science and Artificial Intelligence*

In this section I wish to explore the later work of Varela, who took an independent path from Maturana in the late seventies. Varela has pursued developments in several areas: the immune system (see Section 2.4), neuroscience, cognitive science, and artificial intelligence (AI). I shall consider the latter three, which are obviously interconnected, and we shall see that Varela maintains similar themes, such as antirepresentationalism, but also espouses a less extreme form of constructivism than Maturana. His work also moves beyond the realm of science to explore connections with Buddhism. This is well documented in *The Embodied Mind* by Varela *et al.* (1991), but I shall not discuss this aspect, both because I have no expertise whatsoever in the field and because I personally find it adds little of value.* In the following sections reference will be made to specific papers, but most of the material is also available in the book.

11.3.1. The Development of Cognitive Science

One place to begin is with Varela's (1992) analysis of the development of cognitive science in order to situate the distinctive nature of his work, which is known as the *embodied* or *enactive* approach to cognition. Varela distinguishes four stages:

1. Foundations (1943–53)
2. Cognitivism–computationalism (1956–   )
3. Emergence–connectionism
4. Emergence–enaction

11.3.1.1. Foundations

The foundations of cognitive science were set with the development of *cybernetics*, which means literally "the science of steering." This was an amazingly fertile period, with such people as Weiner (1948), McCulloch (McCulloch and Pitts, 1943), Von Neumann (1958), and Turing (Hodges, 1983), which laid the foundations for computers, artificial intelligence, and cognitivism. The central underlying idea was that thinking—cognition—could be explored *and explained* on the basis of mechanisms, logic, and mathematics. In other words, the brain could be regarded as a machine that worked on logic and that, ultimately, the mechanism of mind[†] was separable from its biological substratum and could be realized in other media, even abstract ones such as a "Turing machine."

---

*These two reasons may well be causally related.
[†]I do not mean to refer to the book by this name by E. de Bono (Penguin, 1979)—

11.3.1.2. Cognitivism–Computationalism

These ideas formed the backbone of the main cognitivist period, which started with meetings among Simon, Chomsky, Minsky, and McCarthy in 1956 and continues, to this day, to be the "normal" paradigm in a Kuhnian sense. Cognitivism forms around the hypothesis that (human) intelligence is, like a computer, *computational* and, in particular, that the brain processes symbols that are related together to form representations of the world outside. Thus, according to this view, cognition occurs by taking in information provided by the environment and forming it into representations, which can then be processed to provide logical responses by way of activity. The metaphor is clearly that of the sequential, humanly programmed computer or information-processing machine.

This computational paradigm has been enormously influential in several areas—psychology, neurobiology, psychoanalysis, and AI—but only the last of these will be presented. For a more detailed discussion see West and Travis (1991a, 1991b). It has formed the basis of mainstream (sometimes known as "formalist") AI, in which the effort has been to produce rationalistic algorithms for performing supposedly intelligent actions. It is based on four principles: that there is a Cartesian separation between mind and body, that thinking consists of manipulating abstract representations, that these manipulations can be expressed in a formal language, and that this is deterministic enough to be embodied in a machine. In practice, this has two requirements: forms of representation and methods of search. Thus, it is first necessary to find some way of formally representing the domain of interest (whether it be, for example, chess, problem-solving, or vision) and then to find some method of sequentially searching the resultant multidimensional space.

AI has been very successful in certain well-defined domains. Probably the best known is chess, where a computer can now play at the grandmaster level. However, cognitive AI systems do not attempt to mimic the way the brain actually works, but try to reformulate situations into something capable of resolution by an efficient search. They assume (following Turing) that the successful performance of "intelligent" actions *is intelligence*, no matter *how* it is performed. In recent years, however, the hegemony of cognitive AI has been breaking down in the face of continued failure in most domains that have been addressed, and attacks have been made on the philosophy of AI (Dreyfus and Dreyfus, 1988; Graubard, 1988; Searle, 1990; Denning, 1990). From expert systems through natural language to robotics, performance at the level of human beings has been the exception rather than the rule. Interestingly, it is the most basic human abilities, such as perception, physical manipulation, and speech that have proved the hardest. Any PC can play a reasonably good game of chess, but ask it to set up the pieces out of the box and you are, literally, in another world. There are currently two responses to these difficulties: *connectionism* and *enaction*.

11.3.1.3. Emergence–Connectionism

One of the main differences between the workings of cognitive AI and the brain is that AI is linear and sequential whereas the brain is parallel and distributed. AI algorithms are simply sequential steps endlessly repeated, but the brain consists of millions of neurons working in parallel. This has two consequences: first, in areas like perception, sequential processing simply cannot do sufficient calculations quickly enough and, second, distributed systems can tolerate localized damage and still function whereas algorithmic processing cannot.

One response to this has been to begin to investigate parallel distributed architectures based on densely connected networks of simple, identical components, a paradigm example being the neural network (Rummelhart and McClelland, 1986; Dorffner, 1993). In fact, this is not new—it was first suggested in the foundation years (Rosenblatt, 1962) but a damning (and ill-informed) report in the early seventies focused attention on cognitivism. Recently, it has come to the fore because of the problems of cognitivism alluded to above, because of improvements in computing power, and because of the growth of dynamic systems and complexity theory (Waldrop, 1992; Kauffman, 1993). These connectionist models work in almost opposite fashion to cognitivist methods and are strongly modeled on the brain. There is no overall controlling center, no prior analysis of the situation, no attempt at formal representation or search. Rather, the network is repeatedly exposed to particular stimuli, and its internal connections are modified in a fashion similar to the modification of neuronal connections. The result is that the network *as a whole* "learns" patterns from its environment. It learns, for example, to distinguish a face in different positions, or ways of conjugating verbs.

There are two main approaches to network learning (Freeman and Skarda, 1988). In one, the network is "trained" by showing it examples that can be classified and then having it adjust its connections (or weights) to minimize the errors made in classification (Rummelhart and McClelland, 1986). In the other approach, it is shown unclassified patterns, and it adjusts its weights to each. After learning, when shown a pattern again it will "recognize" it, even if distortion is present (Hopfield, 1982). These systems are powerful precisely in the areas that cognitive AI is not. They work on the basis not of *symbols*, but of *emergence*. Meaning, in this context, is not a function of any particular symbols, nor can it be localized in particular parts of the network. Indeed, symbolic representation disappears completely—the productive power is embodied within the network structure, as a result of its particular history. Leydesdorff (1993) has produced an interesting comparison between Luhmann's autopoietic theory of society (see Section 8.5) and distributed neural networks.

While connectionism overcomes the problems of sequential search in cognitive AI, Varela argues that it still remains basically representational, even though the representations may be implicit rather than explicit. In essence, it still assumes that there is a pre-given, independent world of objectively definable problem situations and that cognition, whether human or artificial, involves successful representations of such a world. This is where Varela, developing earlier ideas (Chapters 5 and 6), argues against such an objectivist view. Rather, the world we experience is *brought forth,* or *enacted,* through the operations of our organizationally closed biological and nervous systems (see also Brooks, 1991; Shanon, 1988, 1991).

11.3.1.4. Emergence–Enaction

First, it should be noted that Skarda (1992) disagrees with the particular boundary drawn by Varela, while agreeing with the general thrust. Skarda argues that the difference between the two types of neural network mentioned above is important: the first type, which are trained to learn defined patterns, are representationalist, but the second type, which generate their own distinctions, are self-organizing and nonrepresentational. Thus, for Skarda, they would be quite compatible with the enactive paradigm.

The enactive paradigm is based not only on Varela's work, but also that of phenomenologists such as Heidegger (who has already been discussed in Section 7.5.2) and Maurice Merleau-Ponty. It will be the subject of the rest of the chapter.

11.3.2. *The Self and Cognition*

Varela (1991a) outlines a multilayered model of the idea of a self, and this forms the underpinning for his development of enaction. The self, or the question of the identity of an organism, is not a single, unified concept. Rather, Varela depicts a number of interconnected levels of development, with the problem of self-constitution arising at each level. These "regional" selves are

1. biological (cellular identity)
2. bodily (immunological identity)
3. cognitive (behavioral identity)
4. sociolinguistic (personal identity)
5. collective (social identity)

Each domain is an emergent level of development—thus, all living organisms have cellular identity; organisms with self-recognizing immune systems have bodily identity; those with nervous systems have cognitive identity; and humans have linguistic (personal) and social identity. Yet, each level faces the same problem—how to organize the continual reassertion of self in the face of a turbulent environment—and does so in essentially the same manner.

Level 1 is clearly the level of autopoiesis, and I will use this level to illustrate the features of self-constitution common to them all. The first point is the fundamental one of the relations between components and the whole, as well as the occurrence of emergence. This relates also to the two fundamental ways of describing a system—internally, by the interactions of its components, and externally, as a whole within an environment. Here we see an essentially dialectical relationship between the localized interactions of components and the global, emergent properties of the whole (see Fig. 11.1).

There is a reciprocal causality between the two. The interaction of cell components, in an embodiment of the autopoietic organization, gives rise to a whole (unity), which is continuously demarcated from its environment. Yet this global characteristic, in turn, specifies or constrains the components to be of certain types (specification), in certain places (constitution), at certain times (order). This is not a teleological causality—it either happens and autopoiesis continues, or it does not.

The second feature, and one of fundamental importance for later developments, is the nature of the relationship between the organism and its environment. First, it is clear that the *organizational closure* of the autopoietic organism does not imply interactional closure; all biological systems depend on energy and particular chemicals from their milieu. But what is important, and has not been obvious in the past, is the way in which the organism itself specifies what, in its environment, is significant for itself. An organism exists within a niche of possible environ-

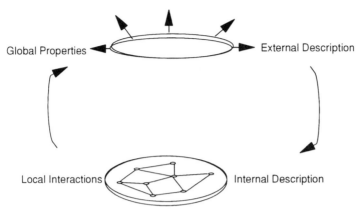

FIGURE 11.1. Relationship between Local and Global Characteristics.

mental interactions of relevance for its continued self-production. It embodies a particular perspective.

In Fig. 11.2, we can see an observer viewing an organism interacting within a context. What we need to distinguish is the context of the organism as seen from the observer's viewpoint (its *environment*) and the context *for* the organism as defined by its structure-determined constitution (see Section 3.1)—its *world*. The structure of the organism determines the possible interactions that it could have and the subset of these that are important for it. For example, we saw in Section 5.3.1 how an

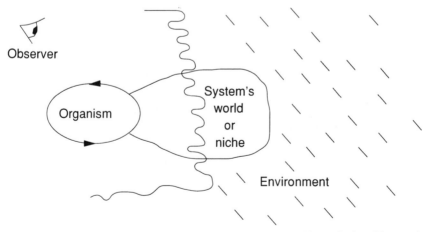

FIGURE 11.2. Relationship among an Organism, Its World, and the Observer's Environment.

amoeba moves toward food through physico-chemical changes in its membrane. That certain substances can have this effect (and others cannot) is determined by the amoeba's internal structure, but that this effect is relevant for the amoeba is determined by its constitution as a whole. Thus the signification or meaning of environmental constituents (e.g., as food or poison) is never absolute, or in itself, but is always relative to a particular organism. Each organism defines its own particular world.

Varela refers to the difference between environment and world as a "surplus of signification" produced by the organism's perspective. The environment in general, unrelated to any particular organism, has no meaning or signification. This is difficult to see as we always, at least implicitly, do refer either to a specific organism, or to ourselves, human beings. But Varela's point is that the environment, in itself, has no meaning. The significance of sucrose as a food is generated entirely by an organism which happens to be able to use it in its ongoing maintenance of identity. Organisms "add meaning" to their environments, and this is an entirely factual matter, determined by their particular internal structure and external unity.

Organisms are always having to undertake activity in order to continue their existence, to replace something that is lacking, to maintain constant some internal relation. This activity, as described by an observer, Varela (and Maturana) refer to as "cognitive" (see Section 5.2), even when it does not involve the nervous system. The motivation for this is to demonstrate the continuity with higher levels, where the term might more usually be applied. This cognitive activity has two aspects— it is a coupling with the environment in order to maintain autopoiesis and, at the same time, the specifying of a world through the generation of signification.

Thus, at the biological level, identity is maintained by a continual interaction with the environment, and this both confirms the organism's demarcation and carves out a world. These same two relations, of part to whole and whole to environment, are also seen at the other levels of self. Note that, so far, this is not new but simply a reformulation of autopoietic theory.

Let us skip the second level, immunological identity, and consider the cognitive identity of the nervous system. The basic characteristics of the nervous system have been described in Chapter 5, so I will summarize these and then describe more recent research. In essence, the nervous system provides an internal coupling between effector and sensory surfaces. It operates as an organizationally closed system, which maintains constant internal relations by compensating for the perturbations it receives, both self-triggered and independently triggered. Its changes of state are determined by its own structure at a time, not by environmental inputs. Most of the activity is, in fact, triggered internally, i.e., by the organism's own bodily movements—all movements trigger sensor neurons as well as being activated by effector neurons. It can be said that *behavior is the control of perception.*

It has become apparent from recent research (outlined in Varela, 1991b) that the nervous system has not developed into simple, clearly differentiated structures with specific one-to-one correspondences (for example, in vision, detectors of position, contrast, movement, and edge), as had been thought. Rather, it has to be seen as a network, or patchwork, of cooperating, but sometimes temporary, subunits. Even quite simple reactions can involve large neuronal groupings, which will change depending on the context:

> It has become evident that these different aspects of vision are emergent properties of concurrent sub-networks, which have a degree of independence and even anatomical separability, but cross-correlate and work together so that a visual percept is this coherency.
> Varela (1991b, pp. 92–93)

We can now consider the constitution of the cognitive self through the two relations discussed for the biological self—the internal relations between part and whole and the external relations with the environment. As with the cell, it is the organizational closure of the nervous system that generates the existence (literally, "standing out") of a self. And the evidence suggests that this is accomplished in a distributed, decentralized manner. Many simple subunits (like Minsky's agents; Minsky, 1987) can come together and give rise to coherent behavior with no central controller or overall top-down structure; they form what Varela calls a "selfless self." "The cognitive self *is* its own implementation: its history and its actions are of one piece" (Varela, 1991b, p. 96).

Moving to the external relation, we find the same situation as with the cell. The nervous system, in a body, has interactions with an environment through both its sensors and effectors and generates a similar (structural) coupling. But, equally, the cognitive self is also situated, and has a perspective, and thus generates a world of signification. It bestows meaning on the events that have significance for it, and environmental events and characteristics become disclosed only in this manner, through the intentionality of the cognitive self. Thus, the cognitive self gains identity only through its situation within a body (its *embodiment*) interacting within an environment, and, as it does so, it constitutes a world of significance in respect to its own self-continuance.

I shall now leave an analysis of the other levels of the self, e.g., the sociolinguistic self, to move directly to Varela's *enactive* view of cognition, a term also used with a different emphasis by Weick (1979).

11.3.3. *The Embodied Mind and Enaction*

The cognitivist program of representationalism, and indeed much of Western philosophy of mind, has been guided by the Cartesian dualism between mind and body—mind being a disembodied realm of pure ideas. However, the thrust of Varela's (and Maturana's) work is in the opposite direction, showing how our cognition and thought is inextricably bound to our *embodied* selves. This is a reversal that has also occurred in phenomenology. We have already seen (Section 7.5) the beginning of this in the difference between Husserl's pure consciousness and Heidegger's concernful, day-to-day, activity. This trend was continued by Merleau-Ponty (1962, 1963), who took phenomenology down to the level of action and perception. Before pursuing this, I should just note that this concern with the body is also occurring in social theory more generally (Shilling, 1993; Synnott, 1993) at this time.

In considering perception, and cognition in general, Merleau-Ponty criticized both what he called empiricist (realist) and intellectualist (idealist) modes of explanation (1962; Hammond *et al.*, 1991). The empiricist simply takes the world as objectively given and sees cognition as a reflection of the world. Intellectualism recognizes that the subject is involved in constituting the experienced world, but is too disembodied and mentalistic. For Merleau-Ponty, cognition is embedded in our body and our nervous system. It is the body which "knows how to act" and "knows how to perceive" on the basis of preformed readinesses and habits developed through its structural coupling with the environment. This is what Varela (1991a) means by "embodiment": the idea that cognition necessarily occurs through and within our bodily structures, which are themselves coupled to biological and social contexts.

Merleau-Ponty also analyzes the relationship between the perceiver and the world, which he sees as a reciprocal relationship—the world does not determine our perception, nor does our perception constitute the world.

> ". . . The properties of the object and the intentions of the subject . . . are not only intermingled; they constitute a new whole." When the eye and the ear follow an animal in flight, it is impossible to say "which started first" in the exchange of stimuli and responses. Since all the movements of the organism are always conditioned by external influences, one can, if one wishes, readily treat behavior as an effect of the milieu. But, in the

same way, since all the stimulations that the organism receives have in turn been possible only by its preceding movements which have culminated in exposing the receptor organ to the external influences, one could also say that the behavior is the first cause of all the stimulations.

> . . . it is the organism itself . . . which chooses the stimuli in the physical world to which it will be sensitive. "The environment *(Umwelt)* emerges from the world through the actualization or the being of the organism— [granted that] an organism can exist only if it succeeds in finding in the world an adequate environment."
> Merleau-Ponty (1963, p. 13), quoted by Varela (1991a, p. 441) (brackets indicate Varela's insertion)

This is the basis of Varela's theory of enactive cognition—that is,

> i) that perception consists in perceptually guided actions;
> ii) that cognitive structures emerge from the recurrent sensory-motor patterns that enable action to be perceptually guided.
> Varela (1991a, p. 441)

The first point is that perception is neither objectivist nor purely constructivist, *pace* Maturana (Varela, 1992, p. 254). Rather, it is codetermined by the linking of the structure of the perceiver and the local situations in which it has to act to maintain its self.

As seen in Fig. 11.3, there can be no fixed point independent of the organism, nor can the organism construct its own closed world. The organism's activity conditions what can be perceived in an environment, and these perceptions, in turn, condition future actions. Varela (1992; Varela *et al.*, 1991, Chapter 8) assembles various neurophysiological evidence for this. For instance, in the area of perception, it is clear that color and smell are by no means simple mappings of external characteristics. Rather, they are cocreations, dependent on the color and smell "spaces" constituted by a particular organism's nervous system and only triggered by external stimulation. Equally, our perception depends for its effectiveness on movement, as shown by Held and Hein's (1958) kittens. Two groups shared the same, artificial, light conditions, but one group were active and the other group passive. When released, the active ones were normal while the passive one acted as if they were blind even though their visual system was unimpaired.

The organism must interact with its environment for its self-continuation, and so the question becomes, how does it happen that the world it carves out is one which permits its continuance? The answer lies not in the world, but in the relations between the sensory and motor surfaces of the nervous system. How is it that these are such as to permit effective, perceptually guided action in a perception-dependent world?

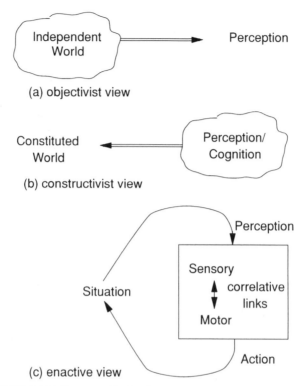

FIGURE 11.3. Perceptually Guided Action: Varela's Enactive View.

This brings us to the second of Varela's points: how action is selected and how the process generates higher cognitive structures.

Our behavior is seen as a constant switching from one task or activity to another according to our readinesses for action. How is it that one is chosen rather than another? This seems to occur as the result of what might be thought of as a competition between different subnets or "agents" in the brain. Brain studies have shown that there are bursts of fast activity followed by more stable patterns as activities stop and new ones start. At each choice-point or breakdown many possibilities are available, but eventually the historically conditioned structure leads to a selection and a new stability. It is next argued that this dynamic interplay linking sensor and motor activity gives rise to the higher cognitive structures. It does not determine them, but does both enable and constrain the more conceptual and abstract modes of thought.

The key here is the emergence of the symbol or sign, and thus language itself (see Section 6.1), as a new domain of neuronal activity. Varela points here to the work of Piaget (1954) and particularly Johnson (1987) and Lakoff (1987; Lakoff and Johnson, 1980). The latter authors are concerned to show how language emerges from and reflects our bodily structure and functioning. For example, much of language can be seen as either direct metaphor or as developing from metaphor, and at base, most metaphors stem from bodily activity. Equally, many of our categorizations, especially the most fundamental ones, arise in relation to our habitual perceptual and motor activities.

We could go further, but the outlines sketched here of Varela's enactive cognition are sufficient.

11.4. Conclusions

So far, the work described in this chapter has not generated as much debate within either the information systems domain or the cognitive science domain as it did in law and family therapy. The most contentious work has been that of Winograd and Flores, which certainly generated much hostility in the AI world when it first appeared. Since then, it has been much referenced and mentioned in a wide variety of fields, but my impression is that it has not yet stimulated a new paradigm with significant new work. It seems to be used as a token gesture by those wishing to distance themselves from traditional positions.

Having said that, however, I think that the arguments put forward in Section 11.3 do show that the representational paradigm in both AI and cognitive science is coming under increasing pressure in the light of its generally accepted failure to produce results. Had *Understanding Computers and Cognition* been published today (1994), it might well have sparked a new wave. I think, therefore, that Varela's work on enactive cognition, developing from autopoiesis and synthesizing connectionism, phenomenology, and current neuroscience, is very exciting and may well provide a framework for future research in the medium term.

IV

Conclusions

Conclusions

Science is not powerful because it is true,
it is true because it is powerful.
Horizon Programme (1986)

12.1. Introduction

We have now reached the end of our exploration of autopoiesis, and it is time to take stock—not by way of an overall judgment and closing of the book, but by simply recalling where we have been and looking toward the future. Section 12.2 provides a brief summary of the main stages of Maturana and Varela's argument, and Section 12.3 reminds us of the principal debates we have encountered.

First, however, let us consider briefly the nature of the work as a whole. How does it stand as a work of science, and how does it fit with Maturana's view of science? The quotation at the start of this chapter can be seen to summarize science for Maturana. Science is not effective and powerful because it is true, because it corresponds to reality; rather, scientific theories become accepted as truth because they are effective, because they are powerful. Cognition is effective action. So autopoiesis is an explanation presented to a community of scientists; it is the proposal of a mechanism which, if it exists, would generate the phenomena experienced by them. Its acceptance as an explanation, and thereby its truth, will depend on its effectiveness, its usefulness, as part of an ongoing conversation among observers.

12.2. Autopoiesis: a Review

In this first section I will review the main points of Maturana's and Varela's work from the basics of autopoiesis to their views on language, cognition, and reality.

12.2.1. Autopoiesis

There is first and of primary importance, as it underlies everything else, the formal specification of the concept of autopoiesis, and how this accounts for the distinction between living and nonliving systems. The term itself does not have any mysterious explanatory power; it merely describes one way in which a set of processes might be connected to one another and claims that, if a system were organized in this way, then it would display the phenomena, such as autonomy and identity despite structural change, that characterize living systems. There is a distinction between the structure and the organization of a system. Structure describes an actual example of a type of system with components and their properties and relations. Organization describes the subset of these relations that determine the system's identity or type: *Autopoietic systems are self-producing systems—their components participate in processes of production the result of which is those very same components.* This has several important implications:

1. The domain of interactions of an organism as a whole, and that of its component parts, are separate and not reducible to each other. An organism interacts locally through its components but this generates, and is constrained by, global properties that are emergent. This does not mean that biological explanations should be, or need be, functionalist or teleonomic. Either autopoiesis continues, or it does not.
2. Autopoietic systems (and some others) are organizationally closed. Organizational closure means that the system (strictly, the system's organization, which determines its identity) is not characterized primarily through inputs and outputs. Rather, all possible states of activity of the system must lead to or generate other, internal states of activity. Other examples of organizationally closed systems are the immune system and the nervous system.

Organizational closure does not imply interactive closure or isolation from the environment. Clearly organisms do, necessarily, interact with their environment. The point is that such interactions also continue the ongoing process of autopoiesis; otherwise, they would not occur. They form part of a circular, self-sustaining process. The result of organizational closure is autonomy—the organization demarcates itself from its environment and, through its own self-referential processes, maintains its self. This also means that the primary domain of biology should be individual organisms (including cells), not species or genes.

3. All composite systems, including autopoietic ones, are structurally determined. This means that the changes that a system undergoes are determined, instant by instant, by its internal structure. They are not determined by the nature of the external action. The environment can act only as a trigger, initiating structural change and co-determining or selecting its path. The resulting changes are a compensation for the interaction.

This has important implications at the cognitive and linguistic levels, since it means that there can be no instructive interactions. Neither symbolic and linguistic tokens nor physical stimuli can impose their meaning; they can only release meaning that already exists structurally.

4. Organisms that interact recurrently with an environment or with other organisms and which have a plastic or changeable structure develop a relationship of structural coupling. That is, particular structural states are selected in the organisms and to some extent in the environment, which are compatible with others. Organisms develop sets of mutually triggering behaviors that are suitable for the maintenance of the organization of each. This is the basis of both adaption to an environment and, at higher levels, linguistic behavior. The resultant path of structural coupling is called ontogenetic structural drift.
5. The nature of structure-dependent systems challenges current notions about the role of information in both biology and psychology. The idea that genes transmit information about the parent organism or that language transmits information from one organism to another are fundamentally mistaken. This is because of the impossibility of instructive interactions. Descriptions such as these may be made by an observer who can see both an organism and its wider environment, but they do not explain the actual operations, which depend only on local, structurally determined interactions.

12.2.2.   *The Nervous System and Cognition*

The analysis of the nature and operation of the nervous system is, in most ways, simply an extension of basic autopoiesis:

1. The first point is how Maturana and Varela use the term *cognition*. For them, the continuation of an autopoietic system implies

or infers that certain necessary interactions will recur in the future. For this reason, they term all autopoietic interactions "cognitive," whether or not the organism has a nervous system. Cognition is effective interaction.

2. The nervous system itself is seen to be organizationally closed. That is, all states of neuronal activity lead to further neuronal activity. Following points 2, 3, and 5 in Section 12.2.1, this opposes the traditional representationalist view of cognition, which would see the environment as generating representations of itself in the brain that would then serve as the basis for cognitive processing in order to arrive at appropriate behavior. In contrast, the main activity of the nervous system is to maintain constant various internal correlations, and the fact that it does allow appropriate behavior is purely a *post hoc* effect of the structural coupling of nervous system to body and body to environment. Perception is not the passive receipt of information but the active construction of a world.

3. The nervous system does, however, allow new domains of interaction to occur. The first domain is that of relations that occur at the sensory surfaces, rather than isolated physical interactions. From this, brain evolution has led to a domain of interactions with internal nervous states, which act as though they were independent. This forms the basis for the emergence of symbolic and linguistic domains and ultimately of language and self-consciousness. All these domains develop because of the evolutionary advantages of cooperation and coordination.

4. As already implied, language is not denotative and representational. That is, words do not carry, in themselves, meaning and reference. Rather, language is arbitrary and consensual—it relies on preestablished agreements, usually taken for granted, developed through structural coupling. Linguistic behavior, or languaging, involves a braiding of language, emotion, and bodyhood. In language we construct the distinctions that constitute our world.

5. Varela has developed a framework for understanding cognition as embodied and enactive. This means that cognition is inseparable from our historically conditioned body and develops through the perceptual control of our actions.

12.2.3. *Philosophical Implications*

Maturana in particular has developed an explicit philosophical position which has become known as radical constructivism:

1. As languaging animals we always exist in language, in the praxis of living. Much linguistic activity consists of explanations of our experiences to ourselves and to others. There are fundamentally two ways of validating explanations—the path of transcendental objectivity and the path of constitutive ontology. In the former, it is assumed that validity comes from some correspondence to an independent reality, and an observer will claim privileged access to this domain. In the latter, it is recognized that we construct our domains of reality through our linguistic behavior, subject to biological constraints. Validity depends on criteria, chosen by the observer, relevant to a particular domain.
2. There are many different nonintersecting domains of reality—domains of viable existence. Each is characterized by its own consistent set of criteria for explanations, valid actions, and operational coherences. Examples are religion, sports, family, science, and the physical domain. They are all equally valid, although not equally desirable.
3. The domain of science is distinctive in being characterized by a particular concern for explanation itself. There is a strong desire for consistency and a particular methodology for validating explanations. Even so, science does not explain an independent reality; it explains our experiences to a community of observers.
4. Following from the above, we can have no access to an independent reality with which to compare our explanations. Indeed, more strongly, no such reality exists, for existence is constituted by the observer. We bring forth the world with our language, and outside of language nothing exists.

12.3. Ongoing Debates

In the preceding chapters we have come across a number of areas of intense debate—what might be called the "active sites" of autopoiesis. In this section I shall briefly recapitulate these as a map of current and future research. I shall deliberately pose them as stark, if somewhat crude, alternatives and shall not recapitulate arguments or conclusions from earlier on.

12.3.1. Noncellular Embodiments of Autopoiesis

Clearly the example *par excellence* of autopoiesis is the biological cell, but there is much debate about other possible examples of autopoiesis—

chemical systems, multicellular systems, and above all, nonbiological systems.

Some chemical systems appear to have some of the characteristics of autopoiesis—e.g., autocatalytic reactions, osmotic growths, and more promisingly, self-replicating micelles, although none are yet fully adequate. Autopoiesis is also important in considering the origin of life and minimal possible cells, with links into Margulis' theory of symbiosis. The situation with regard to multicellular (or metacellular) organisms is in need of major clarification. Do multicellular unities constitute second-order autopoietic systems in their own right, and if so, what exactly are their components—cells, organs, or molecules? Or are second-order autopoietic systems merely the environment or medium for first-order systems but not themselves autopoietic?

Most interestingly, can there be nonphysical autopoietic systems and, if so, would we have to call them "living"? On the first question, Maturana is clear that there can be. There is no restriction in the definition of autopoiesis to the physical domain, and Maturana has said that his computer model is autopoietic. Whether other systems such as law or societies are autopoietic is of course much more problematic and will be discussed below. The second question is more open. Given that we accept an example of nonphysical autopoiesis, such as the computer model, should we call it living? Or should we make living systems a subset of autopoietic systems, or, indeed, restrict autopoietic systems to physical ones?

12.3.2. *Representational versus Embodied Cognition*

There is, here, an enormous fault line, which stretches across all the activity based on the nervous system—perception, cognition, and language itself. Put very crudely, there is the representationalist view, which sees the brain as a system open to the environment—receiving information, processing it in the form of representations, and acting accordingly; which sees language as a denotative and reasonably transparent form of representation; and which sees cognition as a mentalistic process separable in principle from its material embodiment.

In contrast, the embodied view sees the nervous system as organizationally closed—essentially regulating its own internal relations and constitutively unable to interact with some independent reality. Perception is seen as a construction rather than a passive mapping of reality; language, as closed and domain-dependent, triggering preestablished correspondences rather than communicating new information; and cognition as inseparable from its embodiment—indeed, cognition is embodied action. This dichotomy is a reflection of a more general divide, which has existed for some time within the human sciences, surfacing in debates about positivism, realism, interpretivism, phenomenology, and so on. Currently, it is to the fore in cognitive science and AI.

12.3.3. The Ontological Debate

Clearly related to the debate about cognition is one about the ontological status of reality. Is there some stratum independent of human existence and, if so, what form of access do we have to it, if any? Or are we inevitably limited to the realities generated by our own language and activity, either individually or collectively? Do we essentially all live in one world, even though we may experience and understand it differently, or do we inhabit a multiverse of nonintersecting and incommensurable worlds?

This debate has raged most fiercely in the field of family therapy, where it is not merely a matter for interested speculation but has real consequences for particular families and individuals.

12.3.4. The Nature of Social Systems and Law

The question here is to what extent is it possible to characterize social systems and organizations as autopoietic and to what extent would this be merely metaphorical. Varela maintains that social systems are not autopoietic but are organizationally closed. Maturana eventually comes to a similar view, seeing social systems as media in which autopoietic systems can interact and become structurally coupled. Such a system consists of particular patterns of behavior forming a closed domain, but it is open to the entry and exit of participants.

Luhmann, however, finds no problem in conceptualizing societies as autopoietic, but they are constituted not by people but by communications. Societies and their component subsystems are networks of communicative events, each communication being triggered by a previous one and leading in turn to another. Different functional subsystems, such as law, define the nature of their own types of communication and so demarcate and close themselves off, organizationally, from the rest of society. People are not part of society but part of its environment.

This radical theory has triggered much debate about the nature of law; less, so far, about social theory itself.

12.4. Politics and Postmodernism

12.4.1. Political Implications

I now wish to discuss two debates that have not arisen previously in this book—the political implications of Maturana's position and possible resonances between autopoiesis and postmodernism. The political question, which is related to the discussion of power in Section 10.4.2, arises in part from a review of *The Tree of Knowledge* by Berman (1989) and a reply by Maturana (1991b), but has, I think, wider ramifications.

Berman's critique suggests that Maturana has ended up in an apolitical, relativist position, accepting all social arrangements as equally legitimate and being unable to criticize oppressive regimes such as that of Pinochet, which forced Maturana himself to leave Chile. At first sight this is strange, as Maturana has always been very explicit about the ethical implications of his work. It comes about because Maturana's views on constituted reality lead to the conclusion that all domains of reality, all "manners of living," are equally legitimate from a biological point of view. Any mode of life that is successful, in the sense of its own continuance, is a legitimate biological form.

This Maturana accepts, but he argues that this does not mean that all such forms are equally desirable. However, we cannot condemn particular constituted realities on scientific grounds or by claiming that we are objectively right—to do so would be equally oppressive. Rather, we must condemn them because of our preferences—because we do not like them—and we must take responsibility for this choice ourselves and not justify it by reference to some external truth.

This is a deep question, which leads us straight to moral philosophy. Are there any (nonbiological) criteria for judging particular social arrangements, or is it purely a matter for the whim or self-interest of the individual? If it comes down to individual preference, then are such preferences purely accidental or are they ordered in some way? And if they are ordered, are they likely to be related to the society in which the person lives. I will not discuss this in detail here, but would argue that it is not quite as simple as Maturana suggests. The critical theory of Jurgen Habermas (1978, 1984) is very relevant here. This suggests first, that there are three species-wide cognitive interests that transcend particular societies. These are interests in control of the physical world, understanding in the social and linguistic world, and self-development and emancipation in our personal world. This third interest justifies our desire to free ourselves from social situations that are oppressive, both physically and mentally. Second, it suggests that most societies have been oppressive and as such distort both our individual understanding of the world and our collective scientific knowledge. Thus individuals are not necessarily in a position to make free and rational choices.

12.4.2. *Autopoiesis and Postmodernism*

I shall, in this section, briefly note some resonances or shared concerns between autopoiesis and postmodernism. I am certainly not claiming that autopoiesis is whole-heartedly postmodern, but that it can be seen as a bridge between modernism and postmodernism. Characterizing postmodernism would require a book in itself, so I shall be most cursory; for more rounded discussions see Best and Kellner (1991) and Rosenau (1992). We can distinguish, first, between postmodernism and postmodernity. The latter, concerning supposed changes in the nature of industrial society—post-Fordist, postindustrial—does not concern us. The former can also be split into theoretical or epistemological postmodernism and cultural postmodernism. Again, the latter—concerning postmodernism in art, architecture, and aesthetics generally—is not of primary importance for us. This leaves us, then, with epistemological or theoretical postmodernism, based particularly on the work of Baudrillard (1983), Derrida (1978), Foucault (1977), and Lyotard (1984). Postmodernism, taken in a strong form, provides the most radical possible critique of modernism. It negates everything from particular theoretical positions through the very idea of having general theories to rationality and logic itself. Table 12.1 highlights a number of its main elements.

TABLE 12.1
Elements of the Postmodern Approach

Concept	Argument
Indeterminacy	The world is fragmented and chaotic. No underlying structures; only the superficial surface. No objective account—multiple realities.
Antirepresentationalism	Both "reality" and schemes of representation are problematic. Symbols have many meanings and reveal more about their production than about reality.
Decentering the subject	Individual agency and conscious intention is less important than the interplay of meaning and representations.
Textuality	Events and actions are all seen as texts to be read and interpreted in multiple ways. Theory is equally a narrative or story.
Antirationalism	Denies legitimacy to categories such as good/bad, true/false; shuns methods, as well as coherence and consistency; values creativity, spontaneity, ambiguity.

First, there is the view that the world is essentially indeterminate—it is fragmented, disordered, paradoxical. There are no underlying mechanisms or structures governing or determining the surface of events; rather, there is only the superficial, the particular, the fleeting appearance. Equally, there is no history, no long *durée* of development and change, no progress, nothing from which we can learn. There is only the here and now, the present, the way things are at this moment and in this place. There is no global society, only local realities.

Second, postmodernism is antirepresentational; it denies the possibility of any correspondence between representations and objects represented. Lash (1990) provides an interesting contrast. He suggests that for (naive) realism neither reality nor representation is problematic; for modernism (particularly in art), forms of representation become a problem focus; but for postmodernism, reality itself becomes a problem. This follows in part from the first point—there is no independent, fixed reality which can be re-presented and in part from the next point—the decentering or re-placing of the subject. In contrast to modernist logocentrism, which assumes that language can represent external truth, postmodernists argue that representations reveal more about the process of their production than they do about independent reality. Representations, or more generally forms of signification, work through internal differences rather than external similarities. All signs and symbols are inescapably indeterminate, embodying multiple and contradictory meanings and undergoing continual change. Postmodernists emphasize interest in the marginal and in differences, rather than in identity and sameness.

Third, postmodernism downplays the importance of the individual—it decenters the subject—both as creator (author) and as observer (reader). Instead, attention is focused on the way that the subject is created or constructed by the linguistic and cultural position that they (author and reader) occupy. Individual agency and conscious will are seen as less important than the interaction of signifiers and differance, and the distinction between subject and object is denied. Linking to the next point, texts are not seen as the conscious productions of authors, the meaning of which is transmitted to the reader. Rather, texts have no intrinsic meaning; they go beyond their authors, who have no authority over them. They are open to the multiple interpretations made by readers, all of which are equally valid.

Fourth, postmodernists stress the foundational nature of language and text (textuality); indeed, social reality (such as it is) comes to be seen as a text and theory as narrative or story. Thus all events and actions can be "read" as texts, and are equally open to multiple interpretations.

Readers "rewrite" the text each time it is read, and reading is not some form of knowledge gathering but textual construction within a world of multiple realities. A modernist theory is just another narrative, not fundamentally different from any other kind of story, and major worldviews, such as those of Hegel and Marx, are "grand metanarratives."

Finally, postmodernism is antirational, denying the methods and categories of logic and rationality itself. Classical oppositions such as good/bad, true/false, correct/incorrect are seen as superficial irrelevancies, false impositions on an unordered world. All forms of prescriptive method are to be abandoned since none have overriding legitimacy and all constrain creativity and feeling. Clarity and coherence are spurned in favor of ambiguity, partiality, and ambivalence—texts should stimulate and open up, never routinize and close down. Even consistency and the avoidance of contradiction is seen as less important than spontaneity and desire.

In what ways can autopoiesis be said to share these characteristics? First, as Zolo (1993) has said, autopoiesis is certainly postempiricist—it recognizes only too well the problems of gaining some kind of empirical grasp of the world. As we have seen, the constructivist view maintains that we cannot access an objective reality and that we construct, through our language, the multiple realities that we experience. In contrast to postmodernism, however, there is a belief in some underlying order—Maturana's methodology involves hypothesizing possible mechanisms or structures that could generate what we experience.

Second, Maturana and Varela have also been explicitly antirepresentational, criticizing the notion of some grand, freestanding structure of concepts representing the nature of the world and the idea that individual cognition is primarily representational. Instead, autopoiesis sees self-generated, self-constituting domains of activity and discourse (explanatory domains) that are autonomous and incommensurable, each constructing its own self-referential differences and distinctions. There is also a recognition of the fundamental importance and nature of language *per se*. Human beings are constituted by and through their languaging and thereby constitute their worlds. But, as in postmodernism, language is not denotative and logocentric, its symbols representing an objective world. Rather, language is connotative, closed, self-contained, and self-referential. Language is always language in use, intimately connected to our feelings and emotions, and also cognitively tied to our physical and embodied self.

Finally, autopoiesis goes some way toward deemphasizing the individual. While it brings out the importance of the observer (as does postmodernism), the possibilities and limitations of observation are very much bound to our general biological and linguistic characteristics. In the domain of science, the observer is seen as essentially interchangable with other members of the scientific community, and in other domains emphasis is placed on the listener, who is free to interpret and accept, or not accept, what is said. Moreover, Luhmann's development of autopoietic communication very much decenters the subject, putting abstract systems of communication to the fore and relegating the individual to the role of environmental disturbance.

We can interpret this relationship between autopoiesis and postmodernism in two ways. On the one hand, we could see autopoiesis as centering around an essentially modernist project—that of developing scientific explanations for our experiences. But it is one which leads, through its own logical development, to conclusions which affirm some of the tenets of postmodernism. On the other hand, we could see Maturana's later work particularly as beginning with essentially postmodern concerns but then underpinning them with a modernist scientific approach. In either case, autopoietic theory seems to have a foot in both camps.

12.5. In Conclusion

What Maturana and Varela have done is enormously important. Accepting always the limitations of the biological observer, they have followed their own methodology and hypothesized a generative mechanism to account for the phenomena of living systems. Through that mechanism—autopoietic organisms in structural coupling with their medium—they have articulated the emergence of new domains of experience, particularly language; the nature of language as a consensual domain; descriptions and self-descriptions; and finally the observer, including self-referentially an explanation of the very theory itself. They thus demonstrate, in a coherent and consistent way, how even our most self-conscious philosophy emerges from the roots of our biological origins.

I will close this book with a quotation from a poem by W. B. Yeats, "Among School Children," which evokes in me the feeling of autopoiesis:

> *O body swayed to music, O brightening glance,*
> *How can we know the dancer from the dance?*

Bibliography

Guide to the Source Literature

The most comprehensive early papers are:

Maturana, H., 1970a, The neurophysiology of cognition, in: *Cognition, a Multiple View*, (P. Garvin, ed.), Spartan Books, New York, pp. 3–23. This set out the basic view of the nature of cognition and the importance of the observer without using the term autopoiesis.

Maturana, H., 1970a, Biology of cognition, Biol. Computer Lab. Research Report 9.0, U. Illinois, Urbana. Reprinted in: *Autopoiesis and Cognition: The Realization of the Living* (H. Maturana and F. Varela, eds.), Reidel, Dordrecht, 1980, pp. 5–58. This is an expanded version of the previous paper.

Maturana, H., and Varela, F. 1980, Autopoiesis: The organization of the living, in: *Autopoiesis and Cognition: The Realization of the Living*, (H. Maturana and F. Varela, eds.), Reidel, Dordrecht, pp. 73–141. This was written in 1973 as an expanded version of a paper (Autopoietic Systems) first written in 1971 and eventually published in 1975. It coins the term *autopoiesis* as a description of physical, living machines. There is a preface by Stafford Beer.

Two single introductory papers are:

Varela, F., Maturana, H., and Uribe, R., 1974, Autopoiesis: The organization of living systems, its characterization and a model *Biosystems* 5:187–196. This includes the computer model of autopoiesis.

Maturana, H., 1975, The organization of the living: A theory of the living organization *Int. J. Man-Machine Studies* 7:313–332.

Other key works are:

Maturana, H., and Varela, F., 1975, Autopoietic systems—a characterization of the living organization, Biol. Computer Lab Research Report 9.4, U. Illinois, Urbana. The 1971 paper—essentially the same as Maturana and Varela (1973).

Maturana, H., 1978, Biology of language: The epistemology of reality, in *Psychology and Biology of Language and Thought: Essays in Honor of Eric Lenneberg* (G. Millar and E. Lenneberg, eds), Academic Press, New York, pp. 27–63. An important middle work, where Maturana develops his ideas on language as a consensual domain and the idea of subject-dependent realities.

Varela, F., 1979, *Principles of Biological Autonomy*, North Holland, New York. A book describing Varela's own path from autopoiesis, concentrating on organizational closure, algebras, and the immune system.

Maturana, H., and F. Varela, 1980, *Autopoiesis and Cognition: The Realization of the Living*, Reidel, Dordrecht.

Maturana, H., and F., Varela, 1987, *The Tree of Knowledge: The Biological Roots of Human Understanding*, Shambala Publications, Boston. A real effort was made to produce a book for the general reader covering the biological bases of their work.

Maturana, H., 1987, The biological foundations of self-consciousness and the physical domain of existence, in: *Physics of Cognitive Processes*, (E. Caianiello, ed.), World Scientific, Singapore, pp. 324–379. A comprehensive review of all Maturana's work.

Maturana, H., 1988, Reality: The search for objectivity or the quest for a compelling argument, *Irish J. Psych.* 9:25–82. A more accessible version of his views on constitutive reality and the multiverse.

Varela, F., Thompson, E., and Rosch, E., 1991, *The Embodied Mind*, MIT Press, Cambridge, MA. An exposition of Varela's recent work on cognition—the enactive or embodied view.

The references in this bibliography cover only work in English. As autopoiesis has been influential in Germany there are many references in German. A good source for these is

Teubner, G., 1993, *Law as an Autopoietic System*, Blackwell, Oxford.

Special Issues of Journals and Books of Papers

There have been a number of special issues of journals devoted to autopoiesis, in particular:

Family Therapy Networker 9(3), May–June 1985.

Irish J. Psych. 9(1), 1988.

Int. J. Gen. Sys. 21(2), 1992.

Cardozo Law Rev. 13(5), 1992.

There are also a number of books worth highlighting of papers specifically devoted to autopoiesis:

Benseler, F., Hejl, P., and Kock, W. (eds.), 1980, *Autopoiesis, Communication, and Society: The Theory of Autopoietic Systems in the Social Sciences*, Campus Verlag, Frankfurt.

Roth, G., and Schwegler, H. (eds.), 1981, *Self-Organizing Systems—an Interdisciplinary Approach.* Campus Verlag, Frankfurt.

Teubner, G. (ed.), 1987a, *Autopoiesis and the Law*, de Gruyter, Berlin.

Teubner, G. and Febbrajo, A. (eds.), 1992, *State, Law and Economy as Autopoietic Systems: Regulation and Autonomy in a New Perspective*, European Yearbook in the Sociology of Law, Giuffre, Milan.

Veld, R. J. in 't, Schapp, L., Termeer, C. and Twist, M., 1991. *Autopoiesis and Configuration Theory: New Approaches to Societal Steering*, Kluwer, Dordrecht.

Zeleny, M. (ed.), 1980, *Autopoiesis, Dissipative Structures and Spontaneous Social Orders*, AAAS Selectec Symposium 55, Westview Press, Boulder.

Zeleny, M. (ed.), 1981, *Autopoiesis: A Theory of Living Organization.* Elsevier-North Holland, New York.

Interviews with Maturana and Varela

There are also a number of interviews with Maturana and Varela:

Johnson, D., 1976, On observing natural systems—an interview with F. Varela, *Co-Evolution Qu.* Summer:26–31.

Simon, R., 1985a, An interview with Humberto Maturana, *Family Therapy Networker* 9(3):36–37, 41–43.

Krull, M., Luhmann, N., and Maturana, H., 1989, Basic concepts of the theory of autopoietic systems, *Systemic Stud.* 1:79–104.

Electronic Information

There are currently a number of electronic resources available through e-mail over Internet. One is an interactive mailing list managed by Kent Palmer. To subscribe, send a message, "subscribe autopoiesis *your e-mail address*" to

majordomo@world.std.com

If there are any problems you can contact Kent Palmer at

palmer@world.std.com

There is also an electronic newsletter called *The Observer* produced by Randall Whitaker, who also maintains other resources such as bibliographies. To subscribe, contact Randall Whitaker at

rwhitaker@falcon.aamrl.wpafb.af.mil or rhwit@cs.umu.se

References

Alberts, B., Bray, D., Lewis, J., Raff, M., Roberts, K., and Watson, J., 1989, *Molecular Biology of the Cell*, 2nd ed., Garland, New York.

Amselek, P., and MacCormick, N., 1991, *Controversies about Law's Ontology*, Edinburgh U. Press, Edinburgh.

Andrew, A. M., 1979, Autopoiesis and self-organization. *J Cyb.* 9:359.

Ashby, W., 1956, *An Introduction to Cybernetics*, Methuen, London.

Austin, J., 1962, *How to Do Things with Words*, Harvard U. Press, Cambridge, MA.

Bachmann, P., Walde, P., Luisi, P., and Lang, J., 1990, Self-replicating reverse micelles and chemical autopoiesis, *J. Am. Chem. Soc.* 112(22):8200–8201.

Bachmann, P., Luisi, P., and Lang, J., 1991a, Self-replicating reverse micelles, *Chimia* 45(9):266–268.

Bachmann, P., Walde, P., Luisi, P., and Lang, J., 1991b, Self-replicating micelles—aqueous micelles and enzymatically driven reactions in reverse micelles, *J. Am. Chem. Soc.* 113(22):8204–8209.

Baert, P., and De Schampheleire, J., 1987, Autopoiesis, self-organization and symbolic interactionism: some convergences, *Kybernetes* 17(1):60–69.

Bailey, K., 1994, *Sociology and the New Systems Theory: Towards a New Synthesis*, State U. N.Y. Press, Albany.

Banaschewski, B., 1977, On G. Spencer Brown's laws of form, *Notre Dame J. of Formal Logic* 18:507–509.

Bateson, G., 1973, *Steps to an Ecology of Mind*, Granada Publishing, St. Albans.

Bateson, G., 1979, *Mind and Nature: A Necessary Unity*, Dutton, New York.

Baudrillard, J., 1983, *Simulations*, Semiotext(e), New York.

Bednarz, J., 1988, Autopoiesis: The organizational closure of social systems, *Sys. Res.* 5(1):57–64.

Beer, S., 1975, Preface, in: Autopoietic Systems, (H. M. Maturana and F. G. Varela, eds.) Biol. Computer Lab. Research Report 9.4, U. Illinois, Urbana. Reprinted in Maturana, H. M., and Varela, F. G., *Autopoiesis and Cognition: The Realization of the Living*, Reidel, Dordrecht.

Beer, S., 1979, *The Heart of the Enterprise*, Wiley, Chichester.

Beer, S., 1981, *The Brain of the Firm*, 2nd ed., Wiley, Chichester.

Benseler, F., Hejl, P., and Kock, W., eds., 1980, *Autopoiesis, Communication, and Society: The Theory of Autopoietic Systems in the Social Sciences*, Campus Verlag, Frankfurt.

Berger, P., and Luckmann, T., 1967, *The Social Construction of Reality*, Allen Lane, London.

Bergson, H., 1911, *Creative Evolution*, Macmillan, London.

Berman, M., 1989. The roots of reality: Maturana and Varela's The Tree of Knowledge, *J. Human. Psych.* 29(2):277–284.

Best, S., and Kellner, D., 1991, *Postmodern Theory*, Guilford Press, New York.

Bhaskar, R., 1978, *A Realist Theory of Science*, 2nd ed., Harvester Press, Sussex.

Bhaskar, R., 1979, *The Possibility of Naturalism*, Harvester Press, Sussex.

Bhaskar, R., 1986, *Scientific Realism and Human Emancipation*, Verso, London.

Bhaskar, R., 1989, *Reclaiming Reality*, Verso, London.

Birch, J., 1991, Re-inventing the already punctured wheel: Reflections on a seminar with Humberto Maturana, *J. Family Therapy* 13:349–373.

Birleson, P., 1988, Family therapy in the 1980s, *Aust. Paediatr. J.* 24:334–336.

Blom, T., 1990, Why autopoiesis, why systems? in: *Self-Steering and Cognition in Complex Systems: Towards a New Cybernetics* (F. Heylighen, E. Rossel, and F. Demeyere, eds.), Gordon Breech, London, pp. 353–372.

Bloom, S. and Susko, R., 1972, Investigations into the sentential calculus with identity. *Notre Dame J. Formal Logic* 13:289–308.

Bobrow, D. and Collins, A., 1975, *Representation and Understanding: Studies in Cognitive Science*, Academic Press, New York.

Boelen, B., 1975, Martin Heidegger as a phenomenologist, in: *Phenomenological Perspectives* (P. Bossert, ed.), Martinus Nijhoff, The Hague, pp. 93–114.

Bogdan, J., 1988, What's all the fuss? *Family Therapy Networker* 12(5):51.

Boulding, K., 1956, General systems theory—the skeleton of science. *Mgt. Sci.* 2(3):197–208.

Bourdieu, P., 1977, *Outline of a Theory of Practice*, Cambridge U. Press, Cambridge.

Bourgine, P., and Varela, F., 1992, Introduction: Towards a practice of autonomous systems, in: *Towards a Practice of Autonomous Systems: Proceedings of the First European Conference on Artificial Life* (F. Varela, and P. Bourgine, eds.), MIT Press, Cambridge, MA, pp. xi–xvii.

Bowers, J., 1991, Time, representation and power/knowledge, *Theory and Psych.* 1(4):543–569.

Boyd, R. N., 1984, The current status of scientific realism, in: *Scientific Realism* (J. Leplin, ed.), U. California Press, Berkeley.

Briggs, J., and Peat, D., 1985, *Looking Glass Universe: The Emerging Science of Wholeness*, Fontana, London.

Brooks, R., 1991, Intelligence without representation, *Artificial Intelligence* 47:139–159.

Brown, C,. 1988, A new interdisciplinary impulse and the anthropology of the 1990s, *Int. Soc. Sci. J.* 40(2):211–220.

Buchler, J., ed, 1955, *Philosophical Writings of Pierce*, Dover, New York.

Buckley, W., ed, 1968, *Modern Systems Research for the Behavioural Scientist*, Aldine, Chicago.

Buhler, K., 1982, The axiomatization of the language sciences, in: *Karl Buhler: Semiotic Foundations of Language Theory* (R. Innis, ed.), Plenum Press, New York.

Bunge, M., 1979, *Ontology II: A World of Systems*, Reidel Dordrecht.

Burghgraeve, P., 1992, Mechanistic explanations and structure determined systems: Maturana and the human sciences, in: *New Perspectives on Cybernetics: Self-organization, Autonomy and Connectionism* (G. van de Vijver, ed.), Kluwer Academic, Dordrecht, pp. 207–217.

Cade, B., 1986, The reality of "reality" (or the "reality" of reality), *Am. J. Family Therapy* 14:49–56.

Callon, M., 1986, Some elements of a sociology of translation: Domestication of the scallops and the fishermen of St. Brieuc Bay, in: *Power, Action and Belief: A New Sociology of Knowledge* (J. Law, ed.), Sociological Review Monograph 32, Routledge and Kegan Paul, London, pp. 196–233.

Campbell, D., and Draper, R., eds., 1985, *Applications of Systemic Family Therapy: The Milan Method*, Grune and Stratton, New York.

Cannon, W., 1939, *The Wisdom of the Body*, Norton, New York.

Checkland, P., 1971, A systems map of the universe, *J. Sys. Eng.* 2(2):

Checkland, P., 1981, *Systems Thinking, Systems Practice*, Wiley, London.

Clancey, W., 1987, "Understanding computers and cognition—a new foundation for design," a review, *Artificial Intelligence* 31(2):232–250.

Cohen, I., 1989, *Structuration Theory: Anthony Giddens and the Constitution of Social Life*, Macmillan, London.

Cottone, R., 1989, The third epistemology: Extending Maturana's structure determinism, *Am. J. Family Therapy* 17(2):99–109.

Coyne, J., 1985, Towards a theory of frames and reframing: The social nature of frames, *J. Marital and Family Therapy* 11:337–344.

Cull, P., and Frank, W., 1979, Flaws of form, *Int. J. of General Sys.* 5:201–211.

Dawkins, R., 1978, *The Selfish Gene*, Granada Publishing, London.

Dell, P., 1982a, Beyond homeostasis: Towards a concept of coherence, *Family Process* 21:407–414.

Dell, P., 1982b, Family theory and the epistemology of Humberto Maturana, *Family Therapy Networker* 6(4):26,39–41.

Dell, P., 1985, Understanding Bateson and Maturana: Towards a biological foundation for the social sciences. *J. Marital and Family Therapy* 11:1–20.

Dell, P., 1987, Maturana's constitutive ontology of the observer, *Psychotherapy*, 24 (3S):462–466.

Dell, P., 1989, Violence and the systemic view: The problem of power, *Family Process* 28(1):1–14.

Delmonte, M., 1989, Existentialism and psychotherapy: A constructivist perspective, *Psychologia* 32:81–90.

Denning, P., 1990, The science of computing: Is thinking computable? *Am. Sci.* 78:100–102.

Derrida, J., 1978, *Writing and Difference*, RKP, London.

Dewey, J., 1931, *Philosophy and Civilization*, Putnam, New York.

Diamond, S., 1992, Autopoiesis in America, *Cardozo Law Review* 13(5):1763–1769.

Dorffner, G., 1993, How connectionism can change AI and the way we think about ourselves, *App. Artificial Intelligence* 7:59–85.

Dretske, F., 1981, *Knowledge and the Flow of Information*, Blackwell, Oxford.

Dreyfus, H., 1991, *Being-in-the-World: A Commentary on Heidegger's Being and Time, Division I*, MIT Press, Cambridge, MA.

Dreyfus, H., and Dreyfus, S., 1988, Making a mind versus modeling the brain: Artificial intelligence back at a branchpoint, in: *The Artificial Intelligence Debate* (S. Graubard, ed.), MIT Press, Cambridge, MA.

Driesch, H., 1908, *The Science and Philosophy of the Organism*, Black, London.

Ducret, J-J., 1987, Can computers be intelligent? Review of "Understanding Computers and Cognition," *New Ideas Psychol.* 5(2):321–327.

Duhem, P., 1954, *The Aim and Structure of Physical Theory*, Princeton University Press, Princeton, NJ.

Dupuy, J., and Varela, F., 1992, Understanding origins: An introduction, in: *Understanding Origins: Contemporary Views on the Origin of Life, Mind and Society* (J. Dupuy and F. Varela, eds.), Kluwer Academic, Dordrecht.

Efran, J., and Lukens, M., 1985, The world according to Humberto Maturana, *Family Therapy Networker*, 9(3):23–28, 72–75.

Efran, J., Lukens, R., and Lukens, M., 1988, Constructivism: What's in it for you? *Family Therapy Networker* 12(5):26–35.

Emmeche, C., 1992, Life as an abstract phenomenon: Is artificial life possible? in: *Towards a Practice of Autonomous Systems: Proceedings of the First European Conference on Artificial Life* (F. Varela and P. Bourgine, eds.), MIT Press, Cambridge, MA.

Espejo, R., 1993, Domains of interaction between a social system and its environment, *Sys. Practice* 6(5):517–539.

Ewald, F., 1987, The law of law, in: *Autopoiesis and the Law* (G. Teubner, ed.), de Gruyter, Berlin, pp. 36–50.

Faucheux, C., and Makridakis, S., 1979, Automation or autonomy in organizational design, *Int. J. General Sys.* 5:213–220.

Feyerabend, P., 1975, *Against Method*, Verso, New York.

Finnis, J., 1980, *Natural Law and Natural Rights*, Oxford U. Press, Oxford.

Fish, V., 1990, Introducing causality and power into family therapy theory: A correction to the systemic paradigm, *J. Marital and Family Therapy* 16(1):21–37.

Fleischaker, G., 1988, Autopoiesis: The status of its system logic, *Biosystems*, 22:37–49.

Fleischaker, G., 1990, Origins of life: An operational definition, *Origins of Life and Evolution of the Biosphere* 20:127–137.

Fleischaker, G., 1991, The myth of the putative 'organism,' *Uroboros* 1(2):23–43.

Fleischaker, G., 1992a, "Are osmotic or social systems autopoietic?": A reply in the negative, *Int. J. General Sys.* 21(2):163–173.

Fleischaker, G., 1992b, It's not mine and it's not a dictum, *Int. J. General Sys.* 21(2):257–258.

Flores, F., and Ludlow, J., 1981, Doing and speaking in the office, in: *DSS: Issues and Challenges* (G. Fick and R. Sprague, eds.), Pergamon Press, London, pp. 95–118.

Flores, F., Graves, M., Hartfield, B., and Winograd, T., 1988, Computer systems and the design of organizational interaction, *ACM Trans. Office Information Systems* 6(2):153–172.

Fontana, W., and Buss, L., 1994, "The arrival of the fittest": Towards a theory of biological organization, *Bull. Math. Biol.* (forthcoming).

Foucault, M., 1977, *Discipline and Punish: The Birth of the Prison*, Penguin, London.

Foucault, M., 1982, The subject and power, in: *Michel Foucault: Beyond Structuralism and Hermeneutics* (H. Dreyfus and P. Rabinow, eds.), Harvester Press, New York.

Freeman, W., and Skarda, C., 1988, Mind/brain science: Neuroscience or philosophy of mind, in: *Festschrift for John R. Searle* (E. LePore and R. van Gulick, eds.), Blackwell, Oxford, pp. 115–127.

Freifelder, D., 1983, *Molecular Biology*, Science Books International, Boston.

Gadamer, H., 1975, *Truth and Method*, Seabury Press, New York.

Gastelaars, M., 1992, Morality materialized: Notes on public prevention as a policy system, *Sys. Practice* 5(4):411–423.

Gergen, K., 1985, The social constructivist movement in modern psychology, *Am. Psychologist* 40(3):266–275.

Geyer, F., and van der Zouwen, J., 1991, Cybernetics and social science, *Kybernetes* 20(6):81–92.

Giddens, A., 1976, *New Rules of Sociological Method*, Hutchinson, London.

Giddens, A., 1979, *Central Problems in Social Theory: Action Structure and Contradiction in Social Analysis*, Macmillan, London.

Giddens, A., 1984, *The Constitution of Society*, Polity Press, Cambridge.

Goguen, J., and Varela, F,. 1979, Systems and distinctions: Duality and complementarity, *Int. J. General Sys.* 5:31–43.

Goldkuhl, G., and Lyytinen, K., 1982, a language action view of information systems, in: *Proc. 3rd Int. Conf. Information Sys.* (M. Ginzberg and M. Ross, eds.), Ann Arbor, pp. 13–29.

Goldkuhl, G., and Lyytinen, K., 1984, Information systems specification as rule reconstruction, in: *Beyond Productivity: Information Systems for Organizational Effectiveness* (T. Bemelmans, ed.), North Holland, Amsterdam, pp. 79–94.

Goldner, V., 1985, Feminism and family therapy, *Family Process* 24(1):31–47.

Goldner, V., 1988, Generation and gender: Normative and covert hierarchies, *Family Process* 27(1):17–31.

Goldner, V., 1991, Feminism and systemic practice: Two critical traditions in transition, *J. Family Therapy* 13:95–104.

Goldner, V,. Penn, P., Scheinberg, M., and Walker, G., 1990, Love and violence: Gender parodoxes in volatile attachments, *Family Process* 29(4):343–364.

Goolishian, H., and Winderman, L., 1988, Constructivism, autopoiesis and problem determined systems, *Irish J. Psych.* 9, 130–143.

Graubard, S., ed., 1988, *The Artificial Intelligence Debate*, MIT Press, Cambridge, MA.

Griffith, J., Griffith, M., and Slovik, L., 1990, Mind-body problems in family therapy: Contrasting first- and second-order cybernetics approaches, *Family Process* 29: 13–28.

Grzegorczyk, C., 1991, Legal system and reality: A discussion of the autopoietic theory of law, in: *Controversies about Law's Ontology* (P. Amselek and N. MacCormick, eds.), Edinburgh University Press, Edinburgh, pp. 121–141.

Gunji, Y., and Kon-no, N., 1991, Artificial life with autonomously emerging boundaries, *App. Math. Computation* 43:271–298.

Gunji, Y., and Nakamura, T., 1991, Time reverse automata patterns generated by Spencer Brown's modulator: Invertability based on autopoiesis, *Biosystems* 25:151–177.

Habermas, J., 1979, *Communication and the Evolution of Society*, Heinemann, London.

Habermas, J., 1984, *The Theory of Communicative Action, Vol. 1: Reason and the Rationalization of Society*, Heinemann, London.

Habermas, J., 1987, *The Theory of Communicative Action, Vol. 2: Lifeworld and System: a Critique of Functionalist Reason,* Polity Press, Oxford.

Habermas, J., 1990, *The Philosophical Discourse of Modernity: Twelve Lectures,* Polity Press, London.

Habermas, J., 1992, *Postmetaphysical Thinking,* Polity Press, Cambridge.

Hadlington, S., 1992, Autopoiesis—'living' micelles, *Chem. in Britain* (January 1992), p. 10.

Haley, J., 1976, Development of a theory: A history of a research project, in: *Double Bind: the Foundation of the Communicational Approach to the Family* (C. Sluzki and D. Randsom, eds.), Grune and Stratton, NY.

Hamlyn, D., 1987, *A History of Western Philosophy,* Viking, New York.

Hammond, M., Howarth, J., and Keat, R., 1991, *Understanding Phenomenology,* Blackwell, Oxford.

Hanson, N., 1969, *Perception and Discovery, an Introduction to Scientific Inquiry,* Freeman Cooper, San Francisco.

Harnden, R., 1990. The languaging of models: The understanding and communication of models with particular reference to Stafford Beer's cybernetic model of organization structure. *Sys. Practice* 3(3):289–302.

Harnden, R., and Mullery, G., 1991, Enabling network systems (ENS), *Sys. Practice* 4(6):579–598.

Harnden, R., and Stringer, R., 1993a, Theseus—a model for global connectivity, in: *Systems Science: Addressing Global Issues* (F. Stowell, D. West, and J. Howell, eds.), Plenum Press, New York.

Harnden, R., and Stringer, R., 1993b, Theseus—the evolution of a hypermedium, *Cybernetics and Sys.* 24:255–280.

Harré, R., 1970, *The Principles of Scientific Thinking,* MacMillan, London.

Harré, R., 1986, *Varieties of Realism,* Blackwell, Oxford.

Hart, H., 1961, *The Concept of Law,* Clarendon Press, Oxford.

Hassard, J., 1993, Postmodernism and organizational analysis: An overview, in: *Postmodernism and Organizations* (J. Hassard and M. Parker, eds.), Sage, London.

Heidegger, M., 1962, *Being and Time,* Blackwell, Oxford.

Heidegger, M., 1977, The question concerning technology, in: *Martin Heidegger: Basic Writings* (D. Krell, ed.), Harper, San Francisco, pp. 287–317. (Published originally in German in 1954)

Held, B., 1990, What's in a name? Some confusions and concerns about constructivism, *J. Marital and Family Therapy* 16(2):179–186.

Held, B., 1991, Constructing constructivism: A reply to Oz, *J. Marital and Family Therapy* 17(2):193–195.

Held, R., and Hein, A., 1958, Adaption of disarranged hand-eye coordination contingent upon re-afferent stimulation, *Perceptual-Motor Skills,* 8:87–90.

Held, B., and Pols, E., 1985a, The confusion about epistemology and "epistemology"—and what to do about it, *Family Process* 24:507–522.

Held, B., and Pols, E., 1985b, Rejoinder: On contradiction, *Family Process* 24:521–524.

Held, B., and Pols, E., 1987a, Dell on Maturana: A real foundation for family therapy? *Psychotherapy* 24(3S):455–461.

Held, B., and Pols, E., 1987b, The philosophy of Dell and Maturana, *Psychotherapy* 24(3S):466–468.

Heller, T., 1987, Accounting for law, in: *Autopoiesis and the Law* (G. Teubner ed.), de Gruyter, Berlin.

Hodges, A., 1983, *Alan Turing—the Enigma,* Burnett Books, London.

Hoffman, L., 1988, A constructivist position for family therapy, *Irish J. Psych.* 9(1):110–129.

Hoffman, L., 1990, Constructing realities: An art of lenses, *Family Process* 29(1);1–12.

Hofstadter, D., 1980, *Gödel, Escher, Bach: An Eternal Golden Braid*, Penguin, Harmondsworth.

Hopfield, J., 1982, Neural networks and physical systems with emergent collective computational abilities, *Proc. Nat. Acad. Sci. (USA)*, 79:2554–2558.

Horizon Programme, 1986, "Science—fiction?" BBC, London.

Hughes, P., and Brecht, G., 1978, *Vicious Circles and Infinity*, Penguin, Harmondsworth.

Huntington, E., 1904, Sets of independent postulates for the algebra of logic, *Tran. Am. Math. Soc.* 5:288–309.

Husserl, E., 1964, *The Idea of Phenomenology*, Martinus Nijhoff, The Hague.

Husserl, E., 1970, *The Crisis of European Sciences and Transcendental Phenomenology*, Northwestern U. Press, Chicago.

Husserl, E., 1977, *Cartesian Meditations*, Martinus Nijhoff, The Hague.

Jackson, D., 1957, The question of family homeostasis. *Psychia. Qu. Supp.* 31, 79–90.

Jacobson, A., 1989, Autopoietic law: The new science of Niklas Luhmann, *Michigan Law Rev.* 87:1647–1689.

Jakobson, R., and Halle, M., 1956, *Fundamentals of Language*, Mouton, The Hague.

James, W., 1948, *Essays in Pragmatism*, Hafner, New York.

James, K., and McIntyre, D., 1983, The reproduction of families: The social role of family therapy, *J. Marital and Family Therapy* 9:119–129.

Jenkins, H., and Asen, K., 1992, Family therapy without the family: A framework for systemic practice, *J. Family Therapy* 14:1–14.

Joerges, C., and Trubek, D., 1989, *Critical Legal Thought: an American–German Debate*, Nomos, Baden-Baden.

Johnson, D., 1976, On observing natural systems—an interview with F. Varela, *Co-Evolution Qu.* Summer:26–31.

Johnson, D., 1989, A Pickwickian tale: Maturana's ontology of the observer, *Scientific Reasoning Research Institute Report 223*, Hasbrouck Laboratory, U. Massachusetts, p. 57.

Johnson, D., 1991a, A pragmatic realist foundation for critical thinking, *Inquiry: Critical Thinking across the Disciplines*, 7(3):23–27.

Johnson, D., 1991b, The illusions of our epoch: Postmodern discourse and epistemological Robinsonades, *Inquiry: Critical Thinking across the Disciplines*, 8(4):6–8,27.

Johnson, D., 1991c, Reclaiming reality: A critique of Maturana's ontology of the observer, *Methodologia* 9:7–31.

Johnson, D., 1992, The final conceit, *Inquiry: Critical Thinking across the Disciplines*, 9(4):8–12.

Johnson, D., 1993a, Language, thought and world: Confronting the postmodern challenge to knowledge, *Inquiry: Critical Thinking across the Disciplines* 10(3): 3–5.

Johnson, D., 1993b, The metaphysics of constructivism, *Cyb. Human Knowing* 1(4): 24–41.

Johnson, M., 1987, *The Body in the Mind: The Bodily Basis of Imagination, Reason, and Meaning*, U. Chicago Press, Chicago.

Jones, E., 1988, The Milan method—*quo vadis*, *J. Family Therapy* 10:325–338.

Jordan, N., 1968, *Themes in Speculative Psychology*, Tavistock, London.

Jumarie, G., 1987, Towards a mathematical theory of autopoiesis, *Cybernetica* 30(3): 59–89.

Kampis, G., and Csanyi, V., 1991, Life, self-reproduction and information: Beyond the machine metaohor, *J. Theor. Biol.* 148:17–32.

Kauffman, L., 1978, Network synthesis and Varela's calculus, *Int. J. General Sys.* 4:179–187.

Kauffman, L., and Solzman, D., 1981, Letter to the editor, *Int. J. General Sys.* 7:253–256.

Kauffman, S., 1993, *The Origins of Order: Self-Organization and Selection in Evolution*, Oxford U. Press, Oxford.

Keat, R., and Urry, J., 1982, *Social Theory as Science*, 2nd ed., RKP, London.

Keeney, B., 1982, What is an epistemology of family therapy? *Family Process* 21:153–168.

Keeney, B., 1983, *Aesthetics of Change*, Guilford Press, New York.

Kelsen, H., 1967, *Pure Theory of Law*, U. California Press, Berkeley.

Kennealy, P., 1987, Talking about autopoiesis: Order from noise? in: *Autopoiesis and the Law* (G. Teubner, ed.), de Gruyter, Berlin.

Kenny, V., and Gardner, G., 1988, Constructions of self-organising systems. *Irish J. Psych.* 9:1–24.

Kensing, F., and Winograd, T., 1991, The language/action approach to design of computer-support for cooperative work: A preliminary study in work mapping, in: *Collaborative Work, Social Communications and Information Systems* (R. Stamper, P. Kerola, R. Lee, and K. Lyytinen, eds.), Elsevier–North Holland, Amsterdam, pp. 311–332.

Kickert, W., 1993, Autopoiesis and the science of (public) administration: Essence, sense and nonsense, *Org. Stud.* 14(2):261–278.

King, M., 1993, The "truth" about autopoiesis, *J. Law and Society* 20(2):218–236.

King, M., and Piper, C., 1990, *How the Law Thinks About Children*, Gower, Aldershot.

Kohout, L., and Pinkava, V., 1980, The algebraic structure of the Spencer Brown and Varela calculi, *Int. J. General Sys.* 6:155–171.

Kosok, M., 1966, The formalization of Hegel's dialectical logic, *Int. Phil. Quart.*, 6(4), 596–631.

Krull, M., Luhmann, N., and Maturana, H., 1989, Basic concepts of the theory of autopoietic systems, *Systemic Stud.* 1:79–104.

Kuhn, T., 1970, *The Structure of Scientific Revolutions*, 2nd ed., U. Chicago Press, Chicago.

Lakoff, G., 1987, *Women, Fire and Dangerous Things: What Categories Reveal about the Mind*, U. Chicago Press, Chicago.

Lakoff, G., and Johnson, M., 1980, *Metaphors We Live By*, U. of Chicago Press, Chicago.

Lambert, D., and Hughes, A., 1988, Keywords and concepts in structuralist and functionalist biology, *J. Theor. Biol.* 133:133–145.

Langton, C., ed., 1988, *Artificial Life*, SFI Studies in the Science of Complexity, Vol. VI, Addison-Wesley, California.

Langton, C., Taylor, C., Farmer, J., and Rasmussen, S., eds, 1991, *Artificial Life II*, SFI Studies in the Sciences of Complexity, Vol. X, Addison-Wesley, California.

Lash, S., 1990, *Sociology of Postmodernism*, Routledge, London.

Leduc, S., 1911, *The Mechanism of Life*, Rebman, London.

Lee, H., 1961, *Symbolic Logic*, Random House, New York.

Lehtinen, E., and Lyytinen, K., 1986, Action based model of information systems. *Inf. Sys.* 11:299–317.

Lempert, R., 1987, The autonomy of law: Two visions compared, in: *Autopoiesis and the Law* (G. Teubner, ed.), de Gruyter, Berlin, pp. 152–190.

Leplin, J., ed., 1984, *Scientific Realism*, U. California Press, Berkeley.

Lewis, C., and Langford, C., 1959, *Symbolic Logic*, Dover, New York.

Leydesdorff, L., 1993, "Structure"/"action" contingencies and the model of parallel distributed processing, *J. Theory Social Behaviour*, 23(1);47–77.

Leyland, M., 1988, An introduction to some of the ideas of Humberto Maturana, *J. Family Therapy* 10:357–374.

Ludewig, K., 1989, 10 + 1 guidelines or guide-questions: An outline of a systemic clinical theory, *Systemic Stud.* 1:13–32.

Luhmann, N., 1982a, *The Differentiation of Society*, Columbia U. Press, New York.

Luhmann, N., 1982b, The world society as a social system, *Int. J. Gen. Sys.* 8, 131–138.

Luhmann, N., 1983, Insistence on systems theory: Perspectives from Germany—an essay, *Social Forces* 61(3):987–998.

Luhmann, N., 1984a, *Soziale Systeme*, Suhrkamp, Frankfurt.

Luhmann, N., 1984b, The self-description of society: Crisis fashion and sociological theory, *Int. J. Comp. Soc.* 25(1-2):59–72.

Luhmann, N., 1985a, *A Sociological Theory of Law*, RKP, London.

Luhmann, N., 1985b, Society, meaning, religion—based on self-reference, *Sociological Anal.* 46(1):5–20.

Luhmann, N., 1985c, The self-reproduction of law and its limits, in: *Dilemmas of Law in the Welfare State* (G. Teubner, ed.), de Gruyter, Berlin.

Luhmann, N., 1986, The autopoiesis of social systems, in: *Sociocybernetic Paradoxes* (F. Geyer and J. van der Zouwen, eds.), SAGE Publications, London.

Luhmann, N., 1987a, The representation of society within society, *Current Soc.* 35(2):101–108.

Luhmann, N., 1987b, The unity of the legal system, in: *Autopoiesis and the Law* (G. Teubner, ed.), de Gruyter, Berlin, pp. 12–35.

Luhmann, N., 1987c, Closure and openness: On reality in the world of law., in: *Autopoiesis and the Law* (G. Teubner, ed.), de Gruyter, Berlin, pp. 335–348.

Luhmann, N., 1989a, Law as a social system, *Northwestern U. Law Rev.* 83(1,2):136–150.

Luhmann, N., 1989b, *Ecological Communication*, Polity Press, Cambridge.

Luhmann, N., 1990a, *Essays in Self-Reference*, Columbia U. Press, New York.

Luhmann, N., 1990b, The cognitive program of constructivism and a reality that remains unknown, in: *Selforganization: Portrait of a Scientific Revolution* (W. Krohn, ed.), Kluwer, Dordrecht.

Luhmann, N., 1992a, The coding of the legal system, in *State, Law, and Economy as Autopoietic Systems: Regulation and Autonomy in a New Perspective* (G. Teubner and A. Febbrajo, eds.), *European Yearbook in the Sociology of Law*, Giuffre, Milan, pp. 145–185.

Luhmann, N., 1992b, Operational closure and structural coupling: The differentiation of the legal system, *Cardozo Law Rev.* 13(5):1419–1441.

Luhmann, N., 1993, Ecological communication: Coping with the unknown, *Sys. Practice*, 6(5):527–540.

Luisi, P., and Varela, F., 1989, Self-replicating micelles—a chemical version of a minimal autopoietic system, *Origins of Life and Evolution of the Biosphere* 19:633–643.

Lyotard, J.-F., 1984, *The Postmodern Condition: A Report on Knowledge*, Manchester U. Press, Manchester.

Lyytinen, K., and Klein, H., 1985, The critical theory of Jürgen Habermas as a basis for a theory of information systems, in: *Research Methods in Information Systems* (E. Mumford, R. Hirschheim, G. Fitzgerald, and T. Wood-Harper, eds.) North-Holland, Amsterdam, pp. 219–236.

Lyytinen, K., Klein, H., and Hirschheim, R., 1991, The effectiveness of office information systems: A social action perspective, *J. Information Sys.* 1:41–60.

MacKinnon, L., and Miller, D., 1987, The new epistemology and the Milan approach: Feminist and sociopolitical considerations, *J. Martial and Family Therapy* 13(2):139–155.

Maddox, J., 1991, Towards synthetic self-replication, *Nature* 354 (5 December):351.

Margulis, L., 1993, *Symbiosis in Cell Evolution*, Freeman, New York.

Maruyama, M., 1963, The second cybernetics: Deviation-amplifying mutual causal processes, *Am. Sci.* 51:164–179.

Marx, K., 1935 (original 1852), The eighteenth Brumaire of Louis Bonaparte, in: *Selected Works, Volume 1* (K. Marx and F. Engels, eds.) Lawrence and Wishart, Moscow.

Maturana, H., 1970a, The neurophysiology of cognition, in: *Cognition, a Multiple View* (P. Garvin, ed.), Spartan Books, New York, pp. 3–23.

Maturana, H., 1970b, *Biology of Cognition*, Biol. Computer Lab. Research Report, 9.0., Univ of Illinois, Urbana. Reprinted in: *Autopoiesis and Cognition: The Realization of the Living* (H. Maturana and F. Varela, eds.), Reidel, Dordrecht, 1980, pp. 5–58.

Maturana, H., 1974, Cognitive strategies, in: *Cybernetics of Cybernetics* (H. Von Foester, ed.), Biological Computer Laboratory, University of Illinois, pp. 457–469. Also (in French) in *L'Unité de l'homme* (E. Morin and M. Piattelli-Palmarini, eds.), Editions de Seuil, Paris.

Maturana, H., 1975a, The organization of the living: A theory of the living organization, *Int. J. Man-Machine Stud.* 7, 313–332.

Maturana, H., 1975b, Communication and representation functions, Biol. Computer Lab. Report #267, University of Illinois, Urbana. Also in *Encyclopédie de la Pléiade*, Série Méthodique, Psychology Volume, (J. Piaget, ed.), Gallimard, Paris.

Maturana, H., 1978a, Biology of language: The epistemology of reality, in: *Psychology and Biology of Language and Thought: Essays in Honour of Eric Lenneberg* (G. Millar and E. Lenneberg, eds.), Academic Press, New York, pp. 27–63.

Maturana, H., 1978b, Cognition, in: *Wahrnehmung und Kommunikation* (P. Hejl, W. Kock, and G. Roth, eds), Peter Lang, Frankfurt, pp. 29–49.

Maturana, H., 1980a, Autopoiesis: Reproduction, heredity and evolution, in: *Autopoiesis Dissipative Structures and Spontaneous Social Orders* (M. Zeleny, ed.), AAAS Selected Symposium 55, Westview Press, Boulder, pp. 45–79.

Maturana, H., 1980b, Man and society, in: *Autopoietic Systems in the Social Sciences* (F. Benseler, P. Hejl, and W. Kock, eds.), Campus Verlag, Frankfurt, pp. 11–31.

Maturana, H., 1981, Autopoiesis, in: *Autopoiesis: A Theory of Living Organization* (M. Zeleny, ed), Elsevier-North Holland, New York, pp. 21–33.

Maturana, H., 1983, What is it to see? *Arch. Biol. Med. Exp.* 16:255–269.

Maturana, H., 1985, Comment by Humberto R. Maturana: The mind is not in the head, *J. Social Biol. Struct.* 8:309–311.

Maturana, H., 1987, The biological foundations of self-consciousness and the physical domain of existence, in: *Physics of Cognitive Processes* (E. Caianiello, ed.), World Scientific, Singapore, pp. 324–379. (This is essentially the same as the following.)

Maturana, H., 1988a. Ontology of observing: The biological foundations of self-consciousness and the physical domain of existence, *Conference Workbook: Texts in Cybernetics*, American Society for Cybernetics Conference, Felton, CA, 18–23 October.

Maturana, H., 1988b, Reality: The search for objectivity or the quest for a compelling argument, *Irish J. Psych.* 9:25–82.

Maturana, H., 1990, Science and daily life: The ontology of scientific explanations, in: *Selforganization: Portrait of a Scientific Revolution* (W. Krohn, G, Kuppers, and H. Nowotny, eds.), Kluwer Academic Publishers, Dordrecht, pp. 12–35.

Maturana, H., 1991a, Response to Jim Birch, *J. Family Therapy* 13:375–393.

Maturana, H., 1991b, Response to Berman's critique of *The Tree of Knowledge, J. Human, Psych.* 31(2):88–97.

Maturana, H., 1992, Cognition and autopoiesis: A brief reflection on the consequences of their understanding, in: *State, Law and Economy as Autopoietic Systems: Regulation and Autonomy in a New Perspective* (G. Teubner and A. Febbrajo, eds.), *European Yearbook in the Sociology of Law*, Giuffre, Milan, pp. 125–142.

Maturana, H., and Guiloff, G., 1980, The quest for the intelligence of intelligence. *J. Social Biol. Struct.* 3:135–148.

Maturana, H., Lettvin, J., McCulloch, S., and Pitts W., 1960, Anatomy and physiology of vision in the frog, *J. Gen. Physiol.* 43:129–175.

Maturana, H., Uribe, G., and Frenk, S., 1968, A biological theory of relativistic colour coding in the primate retina. *Arch. Biol. Med. Exp.*, Suplemento 1:1–30.

Maturana, H., and Varela, F., 1973, Autopoiesis: The organization of the living, in: *Autopoiesis and Cognition: The Realization of the Living* (H. Maturana and F. Varela, eds.), Reidel, Dordrecht, 1980, pp. 63–134.

Maturana, H., and Varela, F., 1975, *Autopoietic Systems: A Characterization of the Living Organization*. Biol. Computer Lab. Research Report 9.4, U. Illinois, Urbana.

Maturana, H., and Varela, F., 1980, *Autopoiesis and Cognition: The Realization of the Living*, Reidel, Dordrecht.

Maturana, H., and Varela, F., 1987, *The Tree of Knowledge*, Shambhala, Boston.

For Maturana, see also Simon (1985a), Mendez (1988), Krull et al. (1989).

Mayntz, R., 1987, *Politische Steuerung und gesellschäftliche Steuerungsprobleme: Anmerkungen zu einem theoretischen Paradigma, Jahrbuch für Staats- und Verwaltungswissenschaft* 1:89–110.

McCulloch, W., and Pitts, W., 1943, A logical calculus of ideas immanent in nervous activity, *Bull. Math. Biophy.* 5:115–133.

Mead, G.H., 1934, *Mind, Self and Society*. U. Chicago Press, Chicago.

Mendez, C., Coddou, F., and Maturana, H., 1988, The bringing forth of pathology, *Irish J. Psych.* 9:144–172.

Mens-Verhulst, J. van, 1991, Perspective of power in therapeutic relationships, *Am. J. Psychotherapy* XLV(2):198–210.

Merleau-Ponty, M., 1962, *Phenomenology of Perception*, RKP, London.

Merleau-Ponty, M., 1963, *The Structure of Behavior*, Beacon Press, Boston.

Meynen, T., 1992, The bringing forth of dialogue: Latour versus Maturana, in: *New Perspectives on Cybernetics: Self-organization, Autonomy and Connectionism* (G. van de Vijver, ed.), Kluwer Academic, Dordrecht, pp. 157–174.

Miller, J., 1978, *Living Systems*. McGraw Hill, New York.

Mingers, J., 1984, Subjectivism and soft systems methodology—a critique. *J App. Sys. Anal.* 11, 85–103.

Mingers, J., 1989a, An introduction to autopoiesis—implications and applications, *Sys. Practice* 2(2):159–180.

Mingers, J., 1989b, An introduction to autopoiesis: A reply to Fenton Robb, *Sys. Practice* 2(3):349–351.

Mingers, J., 1990, The philosophical implications of Maturana's cognitive theories, *Sys. Practice.* 3(6):569–584.

Mingers, J., 1991, The cognitive theories of Maturana and Varela, *Sys. Practice* 4(4):319–338.

Mingers, J., 1992a, Critiquing the phenomenological critique—autopoiesis and critical realism, *Sys. Practice* 5(2):173–180.

Mingers, J., 1992b, The problems of social autopoiesis, *Int. J. Gen. Sys.* 21(2):229–236.

Mingers, J., 1992c, Technical, practical and critical OR—past, present and future?, in: *Critical Management Studies* (M. Alvesson and H. Willmott, eds.), Sage, London, pp. 90–112.

Mingers, J., 1993a, Information and meaning: Foundations for an intersubjective account, *Warwick Business School Research Papers 74*, U. Warwick, Coventry.

Mingers, J., 1993b, Information and meaning: Towards intersubjective meaning systems, *Warwick Business School Research Papers 75*, U. Warwick, Coventry.

Minsky, M., 1987, *The Society of Mind*, Simon and Schuster, New York.

Monod, J., 1974, *Chance and Necessity*, Fontana, London.

Morgan, G., 1986, *Images of Organization*, Sage, California.

Munch, R., 1992, Autopoiesis by definition, *Cardozo Law Rev.* 13(5):1463–1471.

Nelkin, D., 1987, Changing paradigms in the sociology of law, in: *Autopoiesis and the Law* (G. Teubner, ed.), de Gruyter, Berlin, pp. 191–216.

Norman, D., ed., 1981, *Perspectives on Cognitive Science*, Ablex, Norwood, Mass.

Orchard, R., 1975, On the laws of form, *Int. J. Gen. Sys.* 2:99–106.

Ost, F., 1987, Between order and disorder: The game of law, in: *Autopoiesis and the Law* (G. Teubner, ed.), de Gruyter, Berlin, pp. 70–96.

Outhwaite, W., 1987, *New Philosophies of Social Science: Realism, Hermeneutics and Critical Theory*, Macmillan, London.

Oz, S., 1991, Letter to the editor, *J. Marital and Family Therapy* 17(2):191–192.

Pask, G., 1976, *Conversation Theory*, Elsevier, New York.

Pask, G., 1981, Organizational closure of potentially conscious systems, in: *Autopoiesis: A Theory of Living Organization*, (M. Zeleny, ed.), Elsevier-North Holland, New York.

Perelberg, R., and Miller, A., 1990, *Gender and Power in Families*, Routledge, London.

Piaget, J., 1954, *The Construction of Reality in the Child*, Basic Books, New York.

Platt, R., 1989, Reflexivity, recursion and social life: Elements for a postmodern sociology, *Soc. Rev.* 37(4):636–667.

Popper, K., 1972, *Objective Knowledge: An Evolutionary Approach*. Oxford U. Press, London.

Power, M., 1994, Constructing the responsible organization: Accounting and environmental representation, in *Ecological Responsibility: Self-Organization in Environmental Law*, (G. Teubner, ed.), Belhaven Press, London.

Priban, I., 1968, Models in medicine, *Sci. J.*, June, reprinted in: *Systems Behaviour*, (J. Beishon and G. Peters, eds.), 1972, Open U. Press, London, pp. 222–230.

Prigogine, I., 1980, *From Being to Becoming: Time and Complexity in the Physical Sciences*, Freeman, San Francisco.

Rapoport, A., 1969, *Operational Philosophy*, International Society for General Semantics, San Francisco.

Raven, P., and Johnson, G., 1991, *Understanding Biology*, 2nd ed. Mosby-Year Book, St. Louis.

Rescher, N., 1969, *Many-Valued Logics*, McGraw-Hill, London.

Robb, F., 1985, Towards a 'better' scientific theory of human organizations. *J. Opl. Res. Soc.* 36, 489.

Robb, F., 1989a, Cybernetics and suprahuman autopoietic systems, *Sys. Practice* 2(1):47–74.

Robb, F., 1989b, The application of autopoiesis to social organizations—a comment on John Mingers' "An introduction to autopoiesis: Implications and applications," *Sys. Practice* 2(3):343–348.

Robb, F., 1989c, The application of autopoiesis to social organizations—a comment on John Mingers' reply, *Sys. Practice* 2(3):353–360.

Robb, F., 1989d, The realisation of supra-human processes: The way ahead for cybernetics, in: *Systems Prospects: The Next Ten Years of Systems Research*, (R. Flood, M. Jackson, and P. Keys, eds.) Plenum Press, New York, pp. 133–140.

Robb, F., 1989e, The limits to human organisation: The emergence of autopoietic systems, in: *Operational Research and the Social Sciences* (M. Jackson, P. Keys, and S. Cropper, eds.), Plenum Press, New York, pp. 247–251.

Robb, F., 1991, Accounting—a virtual autopoietic system? *Sys. Practice* 4(3):215–235.

Robb, F., 1992a, Are institutions entities of a natural kind? A consideration of the outlook for mankind, in *Cybernetics and Applied Systems* (C. Negoita, ed.), Marcel Dekker, New York, pp. 149–162.

Robb, F., 1992b, Autopoiesis and supra-human systems, *Int. J. Gen. Sys.* 21(2):197–206.

Rose, S., 1970, *The Chemistry of Life.* Pelican, London.

Resenau, P., 1992, *Post-Modernism and the Social Sciences: Insights, Inroads and Intrusions,* Princeton U. Press, Princeton, NJ.

Rosenblatt, F., 1962, *Principles of Neurodynamics: Perceptrons and the Theory of Brain Dynamics,* Spartan Books, New York.

Rosenfeld, M., 1992, Autopoiesis and justice, *Cardozo Law Rev.* 13(5):1681–1712.

Rosseel, E., and van der Linden, G., 1990, Self-monitoring and self-steering in social interaction: Theoretical comments and an empirical investigation, *Kybernetes* 19(1):18–33.

Roth, G., and Schwegler, H., eds., 1981, *Self-Organising Systems—an Interdisciplinary Approach.* Campus Verlag, Frankfurt.

Rottleuthner, H., 1987, Biological metaphors in legal thought, in: *Autopoiesis and the Law* (G. Teubner, ed.), de Gruyter, Berlin, pp. 97–127.

Rumelhart, D., and McClelland, J., eds., 1986, *Parallel Distributed Processing: Studies on the Microstruture of Cognition* MIT Press, Cambridge, MA.

Russell, B., and Whitehead, A., 1927, *Principia Mathematica,* Cambridge U. Press, Cambridge.

Saussure, F., 1960, *Course in General Linguistics,* Peter Owen, London.

Schutz, A., 1967, *The Phenomenology of the Social World,* Northwestern U. Press, Evanston, Ill.

Schwartz, D., 1981, Isomorphisms of Spencer Brown's laws of form and Varela's calculus for self-reference, *Int. J. of Gen. Sys.* 6:239–255.

Searle, J., 1969, *Speech Acts,* Cambridge U. Press, Cambridge.

Searle, J., 1990, Is the brain's mind a computer program? *Sci. Am.* 262(1):26–31.

Segal, L., 1986, *The Dream of Reality: Heinz von Foerster's Constructivism,* Norton, New York.

Selvini-Palazzoli, M., 1986, Towards a general model of psychotic family games, *J. Marital and Family Therapy* 12:339–349.

Selvini-Palazzoli, M., Boscolo, L., Cecchin, G., and Prata, G., 1978, *Paradox and Counter-Paradox,* Aronsen, New York.

Shanon, B., 1988, Semantic representation of meaning: A critique, *Psych. Bull.* 104(2):70–83.

Shanon, B., 1991, The representational view of mind in perspective: A response to Lucas, *Psych. Bull.,* 110(2):264–267.

Shaw, B., 1925, *Man and Superman. A Comedy and a Philosophy,* Constable & Co., London.

Sheffer, H., 1913, Five independent postulates for Boolean algebra, *Trans. Am. Math. Soc.* 14:481–488.

Shilling, C., 1993, *The Body and Social Theory,* Sage, London.

Simon, R., 1985a, An interview with Humberto Maturana, *Family Therapy Networker,* 9(3):36–37,41–43.

Simon, R., 1985b, A frog's eye view of the world, *Family Therapy Networker,* 9(3):32, 34–35.

Skarda, C., 1992, Perception, connectionism, and cognitive science, in: *Understanding Origins: Contemporary Views on the Origin of Life, Mind and Society* (F. Varela and J. Dupuy, eds.), Kluwer Academic, Dordrecht, pp. 265–271.

Sluzki, C., 1988, Case commentary II, *Family Therapy Networker* 12(5):79–81.

Smith, B., 1991, The owl and the electric encyclopedia, *Artificial Intelligence* 47:251–288.

Smith, F., 1970, Being and subjectivity: Heidegger and Husserl, in: *Phenomenology in Perspective* (F. Smith, ed.), Martinus Nifhoff, The Hague, pp. 122–156.

Speed, B., 1984, How really real is real? *Family Process* 23:511–520.

Speed, B., 1991, Reality exists OK? An argument against constructivism and social constructionism, *Family Therapy* 13:395–409.

Spencer-Brown, G., 1972, *Laws of Form*, Julien Press, New York.

Stefik, M., and Bobrow, D., 1987, "Understanding computers and cognition—a new foundation for design," a review, *Artificial Intelligence* 31(2):220–226.

Steier, F., 1987, "Understanding computers and cognition—a new foundation for design," a review, *J. Commun.* 37(1):135–140.

Stephens, R., and Wood, J., 1991, Information systems as linguistic systems: A constructivist perspective, in: *Systems Thinking in Europe* (M. Jackson, G. Mansell, R. Flood, R. Blackham, and S. Probert, eds.), Plenum, London, pp. 469–474.

Stringer, R., 1992, Theseus, *Int. Fed. Lib. Ass. J.* 18:267–273.

Strong, G., 1988, "Understanding computers and cognition—a new foundation for design," a review, *Behavioural Sci.* 33:77–79.

Suchman, L., 1987, "Understanding computers and cognition—a new foundation for design," a review, *Artificial Intelligence* 31(2):227–232.

Synnott, A., 1993, *The Body Social: Symbolism, Self and Society*, Routledge, London.

Tarbary, J., 1991, Hierarchy and autonomy, *Int. J. Gen. Sys.* 18:241–250.

Taggert, M., 1985, The feminist critique in epistemological perspective: Questions of context in family therapy. *J. Marital and Family Therapy* 11(2):113–126.

Teubner, G., 1984, Autopoiesis in law and society: A rejoinder to Blankenberg, *Law and Society Rev.* 18(2):291–301.

Teubner, G., 1985, After legal instrumentalism? Strategic models of post-regulatory law, in: *Dilemmas of Law in the Welfare State* (G. Teubner, ed.), de Gruyter, Berlin.

Teubner, G., ed., 1987a, *Autopoiesis and the Law*, de Gruyter, Berlin.

Teubner, G., 1987b, Introduction to autopoietic law, in: *Autopoiesis and the Law* (G. Teubner, ed.), de Gruyter, Berlin, pp. 1–11.

Teubner, G., 1987c, Evolution of autopoietic law, in: *Autopoiesis and the Law* (G. Teubner, ed.), de Gruyter, Berlin, pp. 217–241.

Teubner, G., 1987d, Juridification: Concepts, aspects, limits, solutions, in: *Juridification of Social Spheres* (G. Teubner, ed.), de Gruyter, Berlin.

Teubner, G., 1989, How the law thinks: Towards a constructivist epistemology of law, *Law and Society Rev.* 23(5):727–757.

Teubner, G., 1990, And God Laughed . . . indeterminacy, self-reference and paradox in law. *EUI Working Paper 88/342*, European University Institute, Florence.

Teubner, G., 1993, *Law as an Autopoietic System*, Blackwell, Oxford.

Teubner, G., and Febbrajo, A., eds., 1992, *State, Law and Economy as Autopoietic Systems: Regulation and Autonomy in a New Perspective*, European Yearbook in the Sociology of Law, Giuffre, Milan.

Thompson, E., Palacios, A., and Varela, F., 1992, Ways of coloring: Comparative color vision as a case study for cognitive science, *Behavioral and Brain Sci.* 15:1–26.

Tomm, K., 1984, One perspective on the Milan systemic approach: Part 1, overview of development, theory and practice, *J. Marital and Family Therapy* 10:113–125.

Tönnies, F., 1955, *Community and Association*, Routledge & Keegan Paul, London.

Touraine, A., 1977, *The Self Production of Society,* U. Chicago Press, Chicago.

Turney, P., 1986, Laws of form and finite automata, *Int. J. Gen. Sys.* 12(4):307–318.

Ulrich, H., and Probst, G., eds., 1984, *Self-Organization and the Management of Social Systems.* Springer, Frankfurt.

van de Vijver, G., 1992, *New Perspectives on Cybernetics: Self-organization, Autonomy and Connectionism,* Kluwer Academic, Dordrecht.

Van Zandt, D., 1992, The breath of life in the law. *Cardozo Law Rev.* 13(5):1745–1761.

Varela, F., 1971, Self-consciousness: Adaption or epiphenomenon? *Studium Generale,* 24, 426.

Varela, F., 1975, A calculus for self-reference, *Int. J. Gen. Sys.* 2:5–24.

Varela, F., 1976, Not one, not two, *CoEvolution Qu.,* Fall:62–67.

Varela, F., 1977a, Circulus fructuosus: Revisiting self-reference as a scientific notion, *Proc. Annual Meeting of Society for General Systems Research,* Denver, Colorado, pp. 116–118.

Varela, F., 1977b, The nervous system as a closed network, *Brain Theory Newslett.* 2: 66–68.

Varela, F., 1977c, On being autonomous: The lessons of natural history for systems theory, in: *Applied Systems Research* (G. Klir, ed.), Plenum Press, New York, pp. 77–85.

Varela, F., 1979a, *Principles of Biological Autonomy,* Elsevier-North Holland, New York.

Varela, F., 1979b, The extended calculus of indications interpreted as a three-valued logic, *Notre Dame J. Formal Logic* 20:141–146.

Varela, F., 1981a, Describing the logic of the living. The adequacy and limitations of the idea of autopoiesis, in: *Autopoiesis: A Theory of the Living Organization* (M. Zeleny, ed.), Elsevier-North Holland, New York, pp. 36–48.

Varela, F., 1981b, Autonomy and autopoiesis, in: *Self-Organising Systems—an Interdisciplinary Approach* (G. Roth and H. Schwegler, eds.), Campus Verlag, Frankfurt, pp. 14–24.

Varela, F., 1984a, Two principles for self-organization, in: *Self-Organization and the Management of Social Systems,* H. Ulrich, and G. Probst, eds. Springer, Frankfurt.

Varela, F., 1984b, Living ways of sense-making: A middle path for neuroscience, in: *Order and Disorder: Proceedings of the Stanford International Symposium* (P. Livingstone, ed.), Amma Libri, Palo Alto, California, pp. 208–224.

Varela, F., 1988, Structural coupling and the origin of meaning in a simple cellular automaton, in: *The Semiotics of Cellular Communications in the Immune System* (E. Secarz, ed.), Springer-Verlag, Berlin, pp. 151–161.

Varela, F., 1989, Reflections on the circulation of concepts between a biology of cognition and systemic family therapy, *Family Process* 28(March):15–24.

Varela, F., 1991a, Making it concrete: Before, during and after breakdowns, *Rev. Int. Psychopathologie,* 4:435–450.

Varela, F., 1991b, Organism: A meshwork of selfless selves, in: *Organism and the Origins of Self* (A. Tauber, ed.), Kluwer, Netherlands, pp. 79–107.

Varela, F., 1992, Whence perceptual meaning? A cartography of current ideas, in: *Understanding Origins: Contemporary Views on the Origin of Life, Mind and Society* (F. Varela and J. Dupuy, eds.), Kluwer Academic, Dordrecht, pp. 235–263.

Varela, F., and Anspach, M., 1991, Immu-knowledge: The process of somatic individuation, in: *Emergence: the New Science of Becoming* (W. Thompson, ed.), Lindisfarne Press, New York, pp. 68–85.

Varela, F., and Bourgine, P., eds., 1992, *Towards a Practice of Autonomous Systems: Proceedings of the First European Conference on Artificial Life,* MIT Press, Cambridge, MA.

Varela, F., and Coutinho, A., 1989, Immune networks: Getting on the real thing, *Res. Immunol* 140:837–845.

Varela, F., and Coutinho, A., 1991, Second generation immune networks, *Immunol. Today* 12(5):159–167.

Varela, F., and Dupuy, J., eds., 1992, *Understanding Origins: Contemporary Views on the Origin of Life, Mind and Society*, Kluwer Academic, Dordrecht.

Varela, F., and Frenk, S., 1987, The organ of form: Towards a theory of biological shape, *J. Soc. and Biol. Structures* 10:73–83.

Varela, F., and Goguen, J., 1978, The arithmetic of closure, *J. Cyb.* 8:291–324.

Varela, F., Maturana, H., and Uribe, R., 1974, Autopoiesis: The organization of living systems, its characterization and a model, *Biosystems* 5(4):187–196.

Varela, F., Coutinho, A., Dupire, B., and Vaz, N., 1988, Cognitive networks: immune, neural, and otherwise, in: *Theoretical Immunology* (A. Perelson, ed.), SFI Series on the Science of Complexity, Addison-Wesley, New Jersey, 359–375.

Varela, F., Thompson, E., and Rosche, E., 1991, *The Embodied Mind*, MIT Press, Cambridge, MA.

For Varela see also Vaz (1978), Goguen (1979), Thompson (1992), Dupuy (1992), and Bourgine (1992).

Vaz N, and Varela, F., 1978, Self and non-sense: An organism-centered approach to immunology, *Med. Hypothesis* 4:231–267.

Veld, R.J. in 't, Schaap, L., Termeer, C., and Twist, M., 1991, *Autopoiesis and Configuration Theory: New Approaches to Societal Steering*, Kluwer, Dordrecht.

Vellino, A., 1987, "Understanding computers and cognition—a new foundation for design," a review, *Artificial Intelligence* 31(2):213–220.

Vickers, G., 1968, *Value Systems and Social Process*, Tavistock, London.

Von Foerster, H., 1984, On constructing a reality, in: *Observing Systems*, 2nd ed., (H. Von Foerster,), ed. Intersystems Publications, CA, pp. 287–309.

Von Glasersfeld, E., 1984, An introduction to radical constructivism, in: *The Invented Reality*, (P. Watzlawick, ed.), Norton, New York.

Von Neumann, J., 1958, *Computers and Brain*, Yale U. Press, New Haven.

Waldrop, L., 1992, *Complexity: The Emerging Science at the Edge of Order and Chaos*, Viking, London.

Watzlawick, P., ed., 1984, *The Invented Reality*, Norton, New York.

Watzlawick, P., Beavin, J., and Jackson, D., 1967, *Pragmatics of Human Communication*, Norton, New York.

Weick, K., 1979, *The Social Psychology of Organizing*, Addison-Wesley, London.

Weiner, N., 1948, *Cybernetics: Or Control and Communication in the Animal and the Machine*, Wiley, New York.

West, D., and Travis, L., 1991a, The computational metaphor and artificial intelligence: A reflective examination of a theoretical falsework, *AI Magazine*, Spring: 64–79.

West, D., and Travis, L., 1991b, From society to landscape: Alternative metaphors for artificial intelligence, *AI Magazine*, Summer:69–83.

Wilden, A., 1977, *System and Structure*, Tavistock, London.

Winch, P., 1958, *The Idea of Social Science*, Routledge and Keegan Paul, London.

Winograd, T., 1987, A language/action perspective on the design of cooperative work, *Human–Computer Interaction* 3:3–30.

Winograd, T., and Flores, F., 1987a, *Understanding Computers and Cognition*, Addison-Wesley, New York.

Winograd, T., and Flores, F., 1987b, "Understanding computers and cognition—a new foundation for design," a response to the reviews, *Artificial Intelligence* 31(2):250–261.

Wittgenstein, L., 1978, *Philosophical Investigations*, Blackwell, Oxford.

Wolfe, A., 1992, Sociological theory in the absence of people: The limits of Luhmann's systems theory, *Cardozo Law Rev.* 13(5):1729–1744.

Woolgar, S., 1988, *Science: The Very Idea*, Tavistock, London.

Wright, L., and Levac, A., 1992, The non-existence of non-complaint families: The influence of Humberto Maturana, *J. Advanced Nursing* 17:913–917.

Wynne, L., McDaniel, S., and Weber, T., 1986, *Systems Consultations: a New Perspective for Family Therapy*, Guilford, London.

Zeleny, M., ed., 1980, *Autopoiesis, Dissipative Structures and Spontaneous Social Orders*, AAAS Selected Symposium 55, Westview Press, Boulder.

Zeleny, M., ed., 1981, *Autopoiesis: A Theory of Living Organization*, Elsevier-North Holland, New York.

Zeleny, M., and Hufford, C., 1992a, The application of autopoiesis in systems analysis: Are autopoietic systems also social systems? *Int. J. Gen. Sys.* 21(2):145–160.

Zeleny, M., and Hufford, C., 1992b, The ordering of the unknown by causing it to order itself, *Int. J. Gen. Sys.* 21(2):239–253.

Zeleny, M., and Pierre, N., 1976, Simulation of self-renewing systems, in: *Evolution and Consciousness* (E. Jantch and C. Waddington, eds.), Addison-Wesley, New York.

Zolo, D., 1992, The epistemological status of the theory of autopoiesis and its applications to the social sciences, in: *State Law and Economy as Autopoietic Systems: Regulation and Autonomy in a New Perspective* (G. Teubner and A. Febbrajo, eds.), *European Yearbook in the Sociology of Law*, Giuffre, Milan.

Name Index

Subject Index